J. Paulo Davim (Ed.)
Green Composites
Advanced Composites

Also of Interest

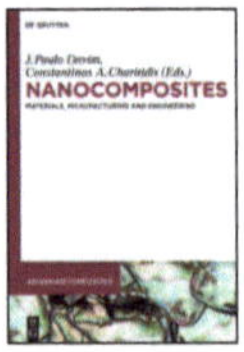

Series: Advanced Composites.
J. Paulo Davim (Ed.)
ISSN 2192-8983
Published titles in this series:
Vol. 6: Wood Composites (2017) Ed. by Aguilera, Alfredo/Davim, J. Paulo
Vol. 5: Ceramic Matrix Composites (2016) Ed. by Davim, J. Paulo
Vol. 4: Machinability of Fibre-Reinforced Plastics (2015)
Ed. by Davim, J. Paulo
Vol. 3: Metal Matrix Composites (2014) Ed. by Davim, J. Paulo
Vol. 2: Biomedical Composites (2013) Ed. by Davim, J. Paulo
Vol. 1: Nanocomposites (2013) Ed. by Davim, J. Paulo/
Charitidis, Constantinos A.

Nanomaterials in Joining.
Charitidis, Constantinos A. (Ed.), 2015
ISBN 978-3-11-033960-4, e-ISBN 978-3-11-033972-7

Nanoparticles.
Jelinek, Raz, 2015
ISBN 978-3-11-033002-1, e-ISBN 978-3-11-033003-8

Progress in Green Tribology.
Green and Conventional Techniques
Davim, J. Paulo (Ed.), 2017
ISBN 978-3-11-037272-4, e-ISBN 978-3-11-036705-8

Green Composites

Materials Manufacturing and Engineering

Edited by
J. Paulo Davim

DE GRUYTER

Editor
Prof. Dr. J. Paulo Davim
University of Aveiro
Department of Mechanical Engineering
Campus Santiago
3810-193 Aveiro, Portugal
pdavim@ua.pt

ISBN 978-3-11-044194-9
e-ISBN (PDF) 978-3-11-043578-8
e-ISBN (EPUB) 978-3-11-043374-6
Set-ISBN 978-3-11-043579-5
ISSN 2192-8983

Library of Congress Cataloging-in-Publication Data
A CIP catalog record for this book has been applied for at the Library of Congress.

Bibliographic information published by the Deutsche Nationalbibliothek
The Deutsche Nationalbibliothek lists this publication in the Deutsche Nationalbibliografie; detailed bibliographic data are available on the Internet at http://dnb.dnb.de.

Cover image: gettyimages/thinkstockphotos, Abalone Shell
Typesetting: PTP-Berlin, Protago-TEX-Production GmbH, Berlin
Printing and binding: CPI books GmbH, Leck
♾ Printed on acid-free paper
Printed in Germany

www.degruyter.com

Preface

Green composites (GCs) are "biocomposites where both matrix and reinforcement derive from renewable resources". GCs are a specific class of biocomposites, where a bio-based polymer matrix is reinforced by natural fibers. They combine plant fibers with natural resins to produce natural composite materials. Recently, GCs have undergone a great development, especially with the introduction of natural fibers from waste. In others words, in general, GCs combine natural fibers with biodegradable resins.

The current volume aims to provide recent information on green composites – materials, manufacturing and engineering – in five chapters. Chapter 1 is dedicated to green composite materials from liquefied biomass. Chapter 2 describes tribological aspects of natural fiber composites. Chapter 3 contains information on investigations on thermal, mechanical and tribological behaviors of hybrid fiber reinforced friction composites. Chapter 4 described machining defects in green composites. Finally, Chapter 5 is dedicated to GUSMRC – from concept to structural application.

The current volume can be used as a research book in a final-year undergraduate engineering course or as a topic on green composites at the postgraduate level. Also, this book can serve as a useful reference for academics, researchers, materials, mechanical and manufacturing engineers, professionals in green composites and related industries. The scientific interest of this book is evident for many important centers of research, laboratories and universities as well as industry. Therefore, it is hoped that this book will inspire and enthuse others to undertake research in this field of green composites.

The editor acknowledges De Gruyter for this opportunity and for their enthusiastic and professional support. Finally, I would like to thank all the chapter authors for their availability for this work.

Aveiro, Portugal, July 2017 J. Paulo Davim

https://doi.org/10.1515/9783110435788-001

Contents

List of contributing authors

Samir Kumar Acharya
Department of Mechanical Engineering
National Institute of Technology
Rourkela, Odisha, India
Chapter 2

M. A. A. Aldahdooh
College of Engineering
Royal University for Women (RUW) &
Western Virginia University (WVU)
P.O. BOX 37400, West Riffa, Kingdom of Bahrain
maged.1987@live.com
Chapter 5

Mehmet Hakki Alma
Department of Forest Products Engineering
Faculty of Forestry
Kahramanmaras Sutcu Imam University
46040, Kahramanmaras, Turkey
alma@ksu.edu.tr
Chapter 1

Ali Alnuaimi
Department of Civil and
Architectural Engineering
Sultan Qaboos University (SQU)
Muscat, Sultanate of Oman
Chapter 5

J.Babu
Department of Mechanical Engineering
St. Joseph's College of Engineering
& Technology
Choondacherry,686579, Kerala, India
jalumedi.babu@gmail.com
Chapter 4

N. Muhamad Bunnori
School of Civil Engineering
University Sains Malaysia (USM)
Malaysia
Chapter 5

Zeki Candan
Department of Forest Products Engineering
Faculty of Forestry
Istanbul University
34473, Istanbul, Turkey
Chapter 1

J.Paulo Davim
Department of Mechanical Engineering
University of Aveiro
Campus Santiago
3810-193 Aveiro, Portugal
Chapter 4

A. Jamrah
College of Engineering
University of Buraimi (UoB)
Al Buraimi, Sultanate of Oman
Chapter 5

D. Lenin Singaravelu
Department of Production Engineering
National Institute of Technology
Tiruchirappalli, 620015, Tamilnadu, India
Chapter 3

M. I. Martini
College of Engineering
University of Buraimi (UoB)
Al Buraimi, Sultanate of Oman
Chapter 5

M. A. Megat Johari
School of Civil Engineering
University Sains Malaysia (USM)
Malaysia
Chapter 5

Suman Kalyan Das
Department of Mechanical Engineering
Jadavpur University
Kolkata, India
Chapter 2

V.Krishnaraj
Department of Production Engineering
PSG College of Technology
Coimbatore, 641004, Tamilnadu, India
vkrishnaraj@hotmail.com
Chapter 3

Jose Philip
Department of Mechanical Engineering
St. Joseph's College of Engineering
& Technology
Choondacherry, 686579, Kerala, India
Chapter 4

S. Manoharan
Department of Mechanical Engineering
SSM College of Engineering
Komarapalayam, 638183, Tamilnadu, India
ma78sce@gmail.com
Chapter 3

Oktay Gonultas
Department of Forest Products Engineering
Faculty of Forestry
Bursa Teknik University
16330, Bursa, Turkey
Chapter 1

Prasanta Sahoo
Department of Mechanical Engineering
Jadavpur University
Kolkata 700032, India
psjume@gmail.com, psahoo@mech.jdvu.ac.in
Chapter 2

B. Suresha
Department of Mechanical Engineering
The National Institute of Engineering
Mysore, 570 008, Karnataka, India
Chapter 3

Tufan Salan
Department of Materials Science
and Engineering
Kahramanmaras Sutcu Imam University
46040, Kahramanmaras, Turkey
Chapter 1

Ayhan Tozluoglu
Department of Forest Products Engineering
Faculty of Forestry
Duzce University
81620, Duzce, Turkey
Chapter 1

R. Vijay
Department of Production Engineering
National Institute of Technology
Tiruchirappalli, 620015, Tamilnadu, India
Chapter 3

Mehmet Hakki Alma, Tufan Salan, Ayhan Tozluoglu,
Oktay Gonultas, and Zeki Candan

1 Green composite materials from liquefied biomass

1.1 Introduction

The energy and chemical needs of the world are continually rising on account of developing technology, industrialization and increasing population. Moreover, the reserves of fossil fuel sources are decreasing rapidly due to these facts and their regeneration cycle time being less than their usage ratio, which makes them nonrenewable resources. Thus, the usage of new and renewable resources as well as eco-friendly fuel and materials technologies have become a good alternative to solving the problems of fossil fuels and their products [1, 2].

Today, lignocellulosic biomass can be converted by thermal, biological or catalytic routes into a great variety of good quality value-added chemicals, biofuels and eco-friendly polymeric materials using a biorefinery system. A special facility called a biorefinery is equipped for the integration of biomass conversion and the production of polymeric materials. A biorefinery can be compared to a modern petroleum refinery in which multiple fuels and petroleum products are produced [3]. Biorefinery technology involves a variety of processes utilizing renewable resources such as grain, lignocellulosic or high moisture content biomass. The product or products are generated leaving no waste and by using every component for conversion in a value-added way supporting ecological sustainability [4].

Materials derived from biomass and particularly wood are among the most plentiful renewable resources on earth. Great effort has been dedicated in recent years to the exploration of techniques for transforming these materials into sustainable products for industrial and commercial usage via the biorefinery concept. One promising way of accomplishing this is through the process of liquefaction; one of the routes of thermochemical platforms of a biorefinery. In the liquefaction process, phenol or multifunctional alcohols are employed to convert wood or such lignocellulosic materials as agricultural and forest residues, waste paper, starch, and the like into low molecular mass liquid products for use in synthesizing polymers [5, 6].

Liquefaction is a promising technology among the thermochemical routes for converting biomass into liquid valuable products via complex chemical and physical reactions. The liquefied wood or wood residues has been considered as a possible bio-alternative raw material for polymer syntheses. In the liquefaction process, basically heating, various catalysts or pressure are used to separate macromolecular substances into small ones. Bio-oils, biocrudes or biopolyols obtained via liquefaction of biomass can be upgraded to gasoline- and diesel-like fuels and various polymeric materials. A significant majority of the main transportation fuels and polymeric biomaterials could

https://doi.org/10.1515/9783110435788-002

be obtained from liquefied biomass, particularly biomass-based wastes, in the near future [7–9].

With the increasing impetus for a more extensive utilization of abundant and renewable biomass resources, liquefaction of wood and its application to green composite materials is a promising technique. Until now, liquefied biomass resins with good processability and reactivity have been successfully prepared by various methods. On the other hand, it has also been found that the liquefaction efficiency of wood, the thermoflow properties and reactivity of the resulting resins were remarkably dependent on the liquefaction conditions such as the kind of catalyst, temperature, etc. Therefore, for a more comprehensive understanding of liquefaction, and for the further development of its applications, more extensive studies on the fundamental phenomena of liquefaction should be accomplished. Thus, this chapter focuses on reviewing the studies about liquefaction of biomass for the production of engineered polymeric materials. Moreover, it explains the principles of the liquefaction methods, properties and synthesis of the liquefied biomass-based materials and pays attention to the importance of biomass-derived materials for the better understanding of the subject.

1.2 Liquefaction technique

In the liquefaction technique, biomass is dissolved in an organic solvent (e.g., phenol, ethylene glycol, glycerol, polyethylene glycol) in the presence of strong (e.g., H_2SO_4) or weak (e.g., $C_2H_2O_4$) acid or alkaline (e.g., NaOH) catalysts at moderate temperatures (80–180 °C) or at elevated temperatures (190–270 °C) without catalyst under atmospheric pressure [10–13].

Liquefaction process of biomass consists of two main steps: dissolution of raw material and filtration of resulting products for the separation of the liquefied part. Fig. 1.1 illustrates the steps of liquefaction and purification of the products. As shown in the figure, after liquefaction, the obtained black liquid is diluted with methanol and filtrated with a glass-fiber filter. The methanol-insoluble part, thus collected, is dried in an oven at 103 ± 2 °C for 24 h and weighed for the determination of the residual part and liquefaction yield. The methanol-soluble fraction is neutralized with MgO or NaOH and filtered with the glass-fiber to remove the salt produced during the neutralization. Subsequently, methanol and free phenol are removed from the methanol-soluble fraction by reduced pressure distillation at 50 °C and 180 °C, respectively. Finally, the liquefied wood resin is ground into powder as a starting material for the synthesis of several polymeric materials [14].

Biomass feedstock naturally comprise polymers such as cellulose, hemicellulose and lignin, which has plenty hydroxyl groups. These hydroxyl groups make it possible to convert liquefied biomass into biopolymers. Liquefaction is an effective way to

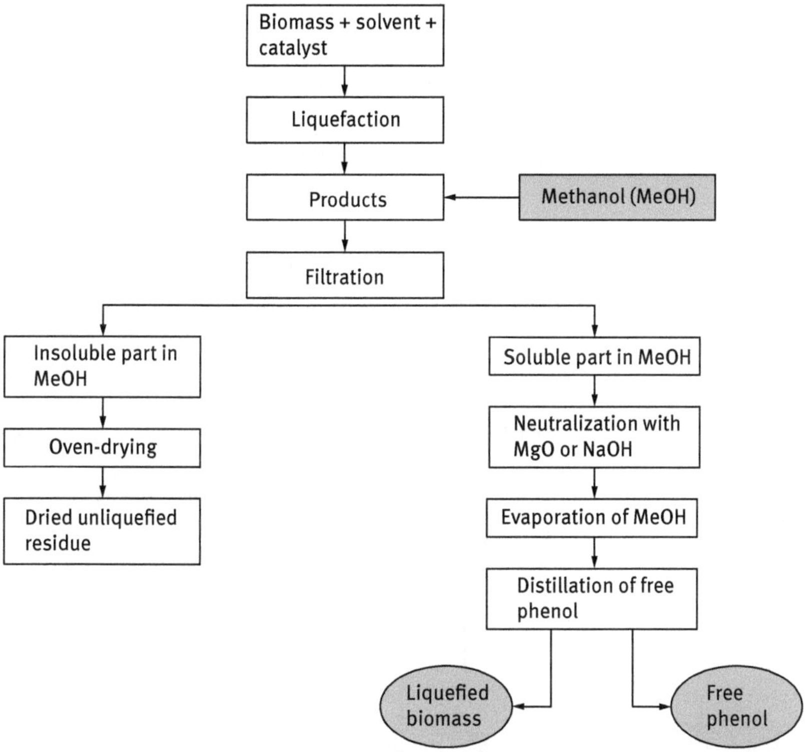

Fig. 1.1: Scheme for the liquefaction procedure of biomass sources.

convert lignocellulosic feedstock into intermediate polyols. Several biopolymeric composite materials can be obtained from these polyols rich in hydroxyl groups [15, 16].

A variety of phenolic compounds such as guaiacol, coniferyl alcohol, vanillin, etc., have been produced due to the cleavage of lignin in the presence of phenol and acid catalyst [17]. Moreover, the cellulose and hemicellulose also undergo transglycosylation to form hydroxymethyl furfural compound in high yield. The furfural thus formed has been found to condense with phenol and formaldehyde through methylene bridges [18].

In the liquefaction process, biomass is converted into liquid products through complex physical and chemical reactions. Polymers of the biomass are broken into small molecules due to various factors such as heat, time, catalyst and solvent. These small molecules are unsteady and reactive, and can polymerize into oil-like substances with a broad range of molecular weight distribution. Different kinds of processes like solvolysis, depolymerization, decarboxylation, hydrogenolysis, and hydrogenation occur during liquefaction phenomena [5].

So far, the pasty biopolyols obtained from liquefaction of biomass have been applied for preparation of novolac [19] and resol-type [12] phenolic resins, molding materials [18], polyurethane foam [20], resol-type phenolic resin foam [21], polyester [15], epoxy resin [22], carbon fibers [23] and activated carbon fibers [24]. Some of the products that can be obtained from liquefied biomass are illustrated in Fig. 1.2.

Fig. 1.2: Scheme of polymeric materials from liquefied lignocellulosics/biomass.

1.3 Foams

Foams are one of the most useful three-dimensional materials with great versatility because they can be used in various forms in several applications areas such as packaging, cushioning and insulation. Polymeric foams include polyurethane foam (PUF), polystyrene foam (PSF) and phenolic foam (PF). The production of these materials consists of varied processing conditions such as gaseous extrusion of molten polystyrene (PS) into foam while the reaction of selected polyols (polyether and polyester) and isocyanate with a blowing agent generate PUF. On the other hand, the PF manufacturing procedure requires the usage of a heat/acid reactive resole-type phenolic resin, emulsifier, a volatile blowing agent and an acid catalyst [25].

The major drawback of these petroleum-based foams is that they are typically manufactured from nonrenewable, nonrecyclable and not-biodegradable raw materials. Accordingly, due to increasing concerns about fossil sources lignocellulosics have become an attractive alternative for the production of bio-based materials.

1.3.1 Polyurethane foams (PUFs) from liquefied lignocellulosics

The various material forms of polyurethanes (PUs) include sheets, foams, elastomers, adhesives, paints, etc., thus making them one of the most versatile three-dimensional polymers. Polyurethanes can be manufactured as a result of the interaction between polyols and polyisocyanate via polyaddition polymerization [26]. Fig. 1.3 shows the main steps of PU-type foam production by using liquefied biomass-based polyol and other additives.

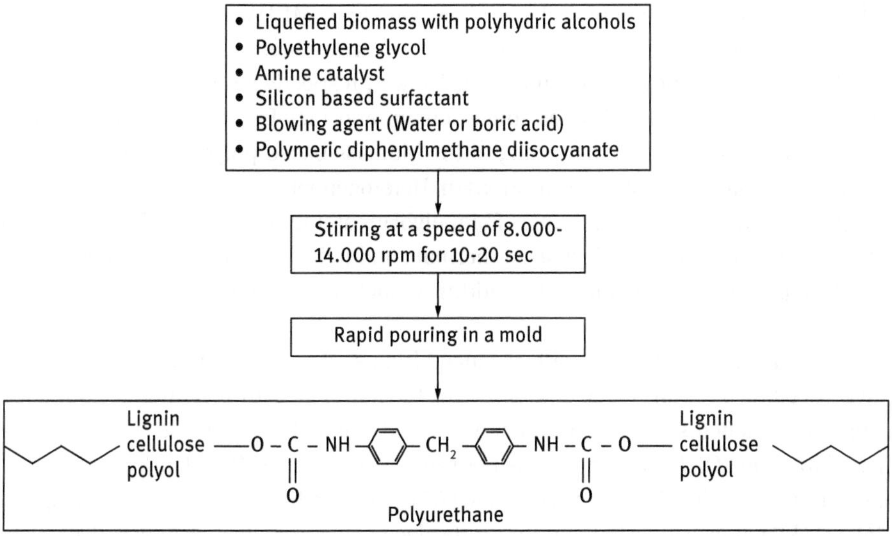

Fig. 1.3: Scheme for the formation of polyurethane from reaction of diphenylmethane diisocyanate (MDI) and natural/synthetic polyols.

It has been proven by many scientists that the liquefied biomass obtained using polyhydric alcohols such as polyethylene glycol (PEG), glycerol or their mixtures as liquefaction solvents could be used directly as polyols for the manufacturing of PUFs without additional treatment. These foams were produced in three different structures comprising the rigid type, using polymeric methylene diphenylene diisocyanate (PMDI) [26]; the semi-rigid type, using polyaryl polymethylene isocyanate (PAPI) [27]; and flexible type using toluene diisocyanate (TDI) [28].

Polyurethane foams are typically produced from polyols with hydroxyl numbers ranging from 300–500 mg KOH/g while biomass typically has hydroxyl numbers around 1500 mg KOH/g. Therefore, the amount of biomass used has to be kept under 33 % to avoid recondensation reactions. The strong hydrogen bonding contained in lignocellulosics is a result of the large number of hydroxyl groups present in the molecular chains of biopolymers. During the liquefaction process, the polyhydric alcohols and acid catalysts lead the disrupting hydrogen bonding with additional hydroxyl groups to struggle with the cellulose inter- and intra-chain hydrogen bonding and their large size impels the chains apart. This phenomenon reduces the efficiency of the process causing a recondensation reaction [29].

To produce polyurethane foams (PUFs), the wood could be liquefed into phenol, bisphenol, alcohols (benzyl alcohol), polyhydric alcohols (1,6-hexanediol, diethylene glycol, etc.), and oxyethers (methyl cellosolve and polyethylene glycol) using acid as a catalyst at a moderate temperature, thus avoiding the use of a catalyst at high temperatures. Rigid PUF was prepared from liquefied sugarcane bagasse biopolyol reacted with commercial methylene diphenyl diisocyanate (p-MDI) and polyethylene glycol (PEG). The researchers used N,N-dimethylcyclohexylamine, water and silicon oil as a catalyst, chemical blowing agent and surfactant, respectively. The results indicated that the prepared PUF exhibited longer cream and tack-free times than the 100 % PEG commercial product. Increasing of the biomass-based polyol content improved the foam density and compressive strength. Heterogeneous surface and irregular pore shape were obtained with amounts of biopolyol greater than 30 %. In addition, the increased biopolyol content decreased thermal conductivity from 0.035 to 0.029 W/mK. The PUF properties were affected by additives such as blowing agents, catalysts, and surfactants [30].

In a further study, PUFs were produced from soybean straw liquefied into crude glycerol using 3 % sulfuric acid at 240 °C for 180 min. The hydroxyl numbers obtained from the liquefied straw ranged from 440 to 540 mg KOH/g. The acid numbers were less than 5 mg KOH/g, and the viscosities ranged from 16 to 45 Pa s. The PUFs thus produced showed a density range of 0.033–0.037 g/cm^3 and had a compressive strength of 148–227 kPa. It was reported that the crude glycerol was an effective alternative polyol for the liquefaction of soybean straw in the making of PU-type foams. The fabricated biopolyols and PUFs resulted in properties comparable to those made with petroleum glycerol-based liquefaction processes [31].

Biodegradable PUF was prepared by liquefying waste paper with polyhydric alcohols (a 4 : 1 w/w mixture of PEG 400 and glycerol) using sulfuric acid as the catalyst. The liquefaction process utilized continuous stirring at 200 rpm along with refluxing at 150 °C for 30–180 min. The liquefied waste paper-based polyol showed suitable chemical and physical properties (e.g., appropriate molecular weight, hydroxyl value, and viscosity) for rigid PUF production. Consequently, it was used effectively with the appropriate mixtures of foaming agents (e.g., water) to produce PUF [32].

The densities and mechanical properties of the foams were acceptable when compared with foams made with polyols based on liquefied wood and starch. The thermal degradation point of the foam was 200 °C and it exhibited similar thermal stability at initial weight losses (0–50 %). In addition, when exposed to leaf mold at 30 °C and 80 % RH for 6 months, the foams showed some degree of degradation, with a weight loss of about 20 %. As a result, they were determined to be potentially biodegradable. Moreover, no mutagens or carcinogens were detected in the water extracted from the foams [32].

Bagasse, which is an important biomass, could be successfully liquefied using PEG# 400 with sulfuric acid. A liquefaction yield of 96 % was achieved by using the polyol, indicating that the lignin was liquefied completely. Being a polyether ester polyol, the liquefied product had the hydroxyl value of ≈ 280–380 mg KOH/g, revealing that the product could be used satisfactorily as a replacement for moderate-intensity PUF [33].

Different wood spices like *Eucalyptus camaldulensis* L. and *Pinus sylvestris* L. were liquefied by employing a polyhydroxy alcohol mixture (PEG# 400 and glycerin) with sulfuric acid at 140–160 °C for 120 min. The dioxane insoluble parts were 17.8 and 5.5 wt% for the liquefied eucalyptus and pine, respectively. It is interesting to note that the lignin of a hardwood, such as eucalyptus, is composed of guaiacyl and syringyl units, while softwood lignin, such as pine, consists of only syringyl units, thus making the hardwood structure more vulnerable to cleavage during the liquefaction process compared to softwood. Highly porous PU-type rigid foams were then fabricated by additionally polymerizing the liquefied woods with polymeric methylene diphenylene diisocyanate (p-MDI) using conventional additives (Fig. 1.4). The compressive strength (120–250 kPa), modulus of elasticity (1–6 MPa), density (0.02–0.05 g/cm^3) and thermal conductivity (0.0352–0.0374 W/mK) of the PUFs prepared from the eucalyptus and pine wood-based foams were found to be nearly comparable to those of commercial PUFs. Moreover, the PU-type foams produced from biomass were more biodegradable than commercial ones [34].

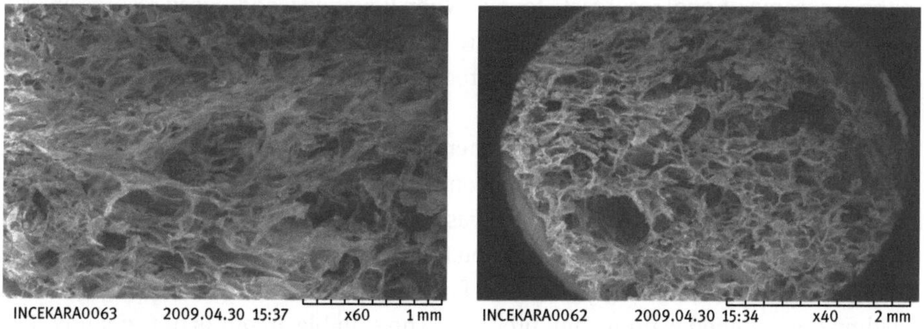

Fig. 1.4: SEM pictures of eucalyptus-based PUF (left) and pine-based PUF (right) [34].

1.3.2 Phenolic foam from liquefied lignocellulosics

Phenolic foams have attracted great attention due to their perfect fire resistance, low fire toxicity, high dimensional stability, and low thermal conductivity compared to other foam types. Liquefied wood was also used for the production of phenolic foam. The wood sample was firstly liquefied with sulfuric acid as the catalyst at a moderate temperature (150 °C) for 150 min under constant stirring and reflux. After the reaction, the mixture was added to methanol, stirred for 20 min and then filtered under a reduced pressure using a glass-fiber filter. A 37 % aqueous solution of formalin was added to the liquefied product, and the pH adjusted by adding sodium hydroxide. The resinification of the liquefied wood was carried out under constant agitation at 55–90 °C and the resin was then cooled to 40 °C and neutralized using sulfuric acid. Finally, pressure-reduced evaporation was used to remove volatile components and concentrate the resin [21].

Using poly(ethylene ether) of sorbitan monopalmitate as the surfactant, the liquefied wood-based resol resin was then mixed vigorously for 10–20 s with a 37 % hydrochloric acid catalyst, while using a diisopropyl ether blowing agent. It was then immediately transferred to an open-top square wooden box and cured for 1 h at room temperature before removal from the box [21].

An excellent yield of liquefied wood-based resol resin was produced by the reaction of liquefied wood and formaldehyde under alkaline conditions. An investigation was carried out on the effects of various reaction parameters on the level of the yield, the unreacted phenol content, and the viscosity of this resin. Compared to conventional methods, this technique involved milder resol resinification conditions. With appropriate mixtures of foaming agents, the liquefied wood-based resol resin was used successfully in the production of phenolic foam. Because of its relatively higher boiling temperature, diisopropyl ether was a suitable foaming agent for the liquefied, wood-based resol resin, with hydrochloric acid and poly(ethylene ether) of sorbitan monopalmitate being employed as the catalyst and surfactant, respectively. The resulting foams exhibited satisfactory densities and compressive properties comparable to those of conventional resol resin foams. When liquefied wood-based resol resin was blended with conventional resol resin, low-density foams were produced [21]. Fig. 1.5 presents the phenolic foam production procedures by using liquefied biomass-based polyols.

In another study, walnut shell was phenolated (phenol/walnut shell ratio of 5 : 1) with sulfuric acid catalyst (3 wt% based on the walnut shell weight) in a three-flask reactor at 120 °C for 30 min. The mixture was kept under constant stirring and reflux in an oil bath. When liquefaction was complete, an equal amount of sodium hydroxide (50 % aqueous solution) was introduced for neutralization of the sulfuric acid, thus yielding the liquefied walnut shell products. The suitable amounts of formalin (37 % aqueous solution) and sodium hydroxide were calculated and added to the liquefied products in order to retain the phenol/formaldehyde/NaOH mol ratio of 1 : 1.8 : 0.02.

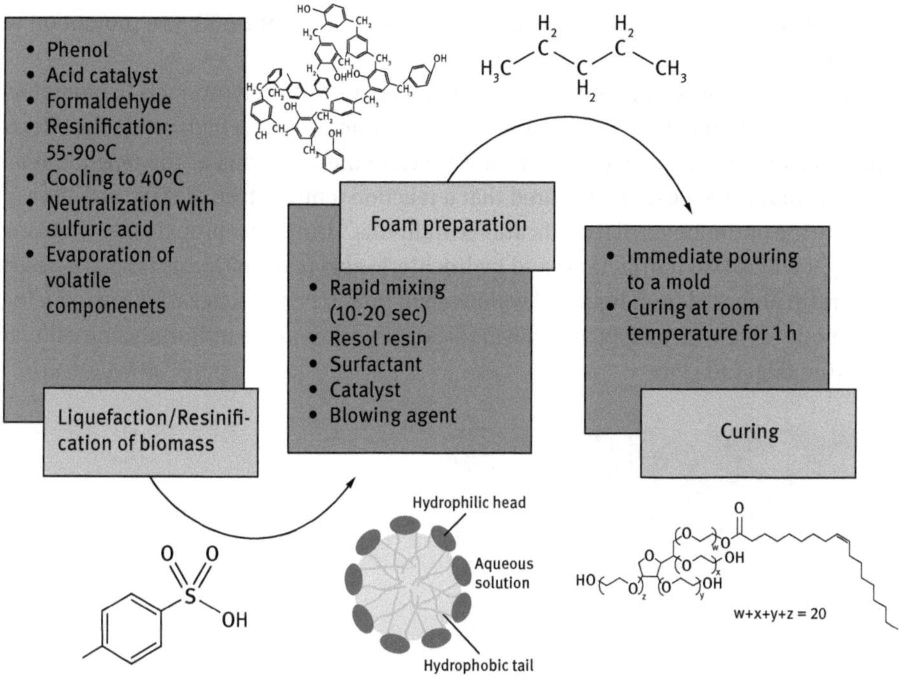

Fig. 1.5: Phenolic foam production from liquefied biomass.

A co-condensation reaction was carried out in 2 h under reflux at 80–82 °C and the resulting resin was then cooled to 40 °C and neutralized by adding hydrochloric acid. Finally, it was subjected to pressure-reduced evaporation in order to remove volatile components and concentrate the resin. A comparison was made by following the standard procedure for synthesis of conventional resol-type phenol formaldehyde resin [35].

The resol resin consisted of about 80 % solids and a mixture of foaming agents was used in the resol-type phenolic foaming. It is recognized that for liquefied wood-based resol resin, diisopropyl ether, because of its 68.4 °C boiling point, is the most suitable for foaming. Therefore, this study employed diisopropyl ether as the blowing agent, which resulted in good foaming and facilitated the formation of uniform cells throughout the foam. The optimum amount of blowing agent was determined as 5–15 %, according to the density of the foam. The surfactant is crucial in the preparation of phenolic foam because it can lessen the surface tension between blowing agent (nonpolar) and the resol resin (polar) [35].

The amount of surfactant was an important consideration in the foaming because the surfactant is responsible for maintaining the balance between foam stability and the collapse of cells. That study used 2–10 % TWEEN® 80 (i.e., polysorbate 80) as a surfactant, which appeared to be a suitable amount as the resulting foam did not collapse. By adding a foaming catalyst of hydrochloric acid at 80 °C, the promotion of resol resin crosslinking was facilitated [35].

This resol-type resin, produced from liquefied walnut shell products in phenol reacted with formaldehyde under low alkaline conditions, gave a high yield. The effects on the yield and viscosity of this resin were examined in terms of the reaction temperature and time. Results indicated that a reaction temperature of 80 °C and time of 2 h were the optimum resol resinification conditions. Using diisopropyl ether (blowing agent), TWEEN®80 (surfactant) and hydrochloric acid (catalyst), this biomass-based resol resin prepared from liquefied walnut shell products was successful in producing a phenolic foam with satisfactory mechanical properties and a uniformly fine cellular structure (Fig. 1.6) [35].

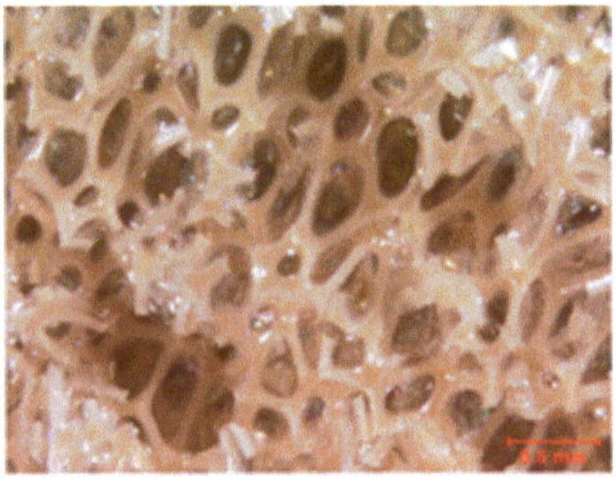

Fig. 1.6: Microphotograph of phenolic foam from biomass-based resol resin [35].

1.4 Molding materials

Liquefaction of biomass sources by using phenol as a reagent solvent and an acid catalyst has long been studied as a technique to utilize biomass as an alternative to petroleum-based products. The effects of water content, catalyst type, catalyst concentration, liquefaction temperature and time, and phenol to raw material ratio have been widely investigated. The molecular weight and flow properties of the liquefied wood have also been characterized [14, 18, 36].

In a very early work, Mcelhinney [37] patented a method for the conversion of different lignocellulosic feedstocks into phenolic resin with an acidic catalyst at high pressure. Moreover, Hesse and Jung [38] patented a wood liquefaction technique in the presence of sulfuric acid for the production of molding and coating resins with hexamethylenetetramine (HMT). Especially in Japan during the last two decades, scientists have conducted many investigations on the biomass liquefaction process into phenol for the production of bakelite-like molding materials [14, 36, 39, 40].

The properties of the phenolated wood, prepared by using acidic catalysts, except for oxalic acid, and the phenolated wood-based molded materials from it were found to be quite comparable to those of commercial novolac resin. The studies showed that among these catalysts, sulfuric acid appeared to be the most effective catalyst on the phenolation of unmodified wood at a moderate temperature [14]. Moreover, NaOH used as a catalyst in the phenolation of wood at elevated temperatures resulted in the good dissolution of wood into phenol, but with large amounts of phenol remaining unreacted. Moreover, the properties of NaOH-catalyzed phenolated wood and molded materials prepared from it, e.g., thermofluidity and mechanical properties, were determined to be evidently lower than those of commercial novolac resin due to insufficient amounts of phenol reacted with wood components [20].

As results of these findings, it was necessary to find a way to diminish the amount of unreacted phenol, which led to many problems when compared with commercial novolac resin. It was found that almost all the lignocellulosic biomass could be successfully liquefied into the solvents. Specifically, phenol formaldehyde-type moldings from the wood liquefied into phenol by using inorganic and organic acidic catalysts had physical and mechanical properties comparable to commercial phenol formaldehyde-type moldings [36]. The novolac resin-type moldings gained much more attention than those of the others. In the production of materials of this type, after being mixed with a curing agent and other conventional molding agents, the resultant phenolated wood composites can be molded at a temperature of 170–200 °C for about 5 min under high pressure [41].

Fig. 1.7 summarizes the preparation procedure for the resinified phenolated biomass and its application to molding materials. As shown in Fig. 1.7, after the phenolation of the biomass at Stage I, the temperature of the phenolated mixture is allowed to decrease to 100 °C. Then, subsequently, at Stage II, a formaldehyde solution, acid, and water are added to the mixture, and then the resinification reaction is conducted for 1.5 h at 100 °C with constant stirring. The resulting mixture thus obtained is diluted with methanol and filtrated with a glass-fiber filter. Biomass residues, the so-called "methanol-insoluble parts", are oven-dried for 24 h and weighed to measure their amounts of methanol-insoluble parts. The soluble parts are first evaporated at 50 °C under a vacuum and then concentrated at 180 °C under a reduced pressure of about 45 mm Hg for 1 h to remove the free phenol, formaldehyde, and water [14].

The oven-dried phenolated biomass is dissolved in acetone and mixed with HMTA (curing agent), zinc stearate (lubricating agent), $Ca(OH)_2$ (accelerating agent) and biomass flour (filler). After mixing thoroughly, the mixture was oven-dried at 70 °C for 1 h to remove acetone, and then ground in a mortar to make a fine powder. For the measurement of mechanical properties, the powder of the phenolated biomass resin so obtained was compression molded into test specimens in a die with of the desired dimensions. The molding conditions were as follows: temperature, 190 °C; pressure, 39.2 MPa; time, 5 min; and cooling, under a slight pressure from 190 °C to ambient temperature [14].

Candan et al. [42, 43] developed carbon nanotube (CNT)-reinforced biocomposites from liquefied biomaterial waste. Dynamic mechanical thermal analysis (DMTA), thermogravimetric analysis (TGA), and scanning electron microscopy (SEM) analysis

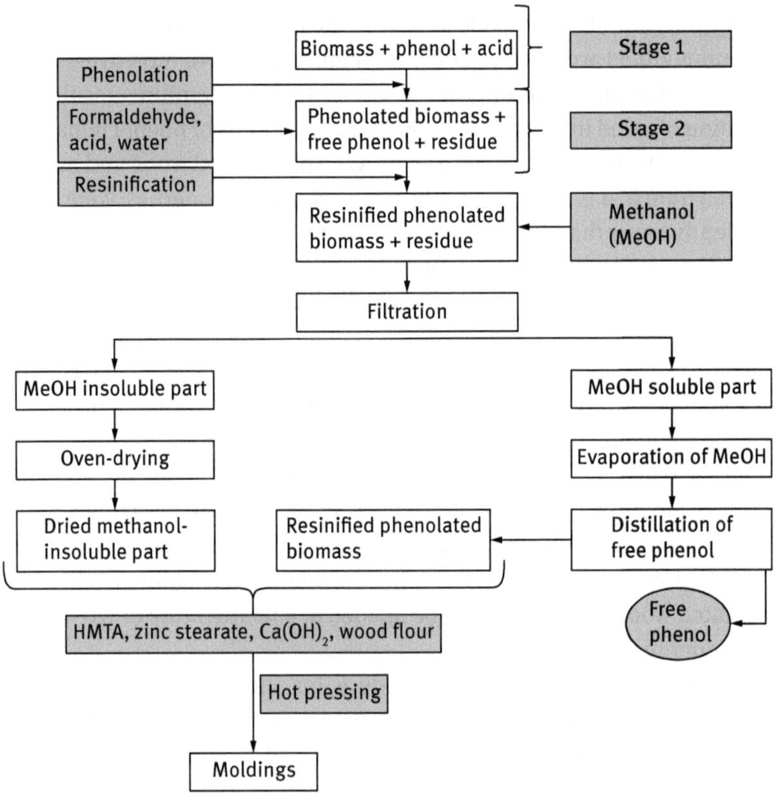

Fig. 1.7: Scheme of the preparation of resinified phenolated biomass and molding production.

of the bio-nanocomposite materials were then performed. It was stated that the storage modulus, loss modulus, and tan delta characteristics of the bio-nanocomposite materials were influenced by the CNTs. It was also reported that the CNTs' reinforcement increased thermal stability of the biocomposites. It was concluded that the bio-nanocomposites having enhanced mechanical and thermal properties might have a potential to be used as a novel biocomposite material for aerospace and automotive applications.

1.4.1 Liquefied wood as replacement in novolac-type resin-based composites

Sulfuric, hydrochloric, phosphoric or oxalic acids can all be utilized as catalysts in the liquefaction of lignocellulosic materials [11, 20, 44–46]. Sulfuric acid appears to be the most efficient catalyst in the phenolation of unmodified wood when moderate temperatures are used. Liquefied wood products with a higher combined phenol and lower wood residue content can be attained with a strong acid rather than with a weak acid as catalyst. Wood component phenolysis using an acidic catalyst can result in a great number of different reactions in competition with each other [47].

Liquefied wood resins resulting from liquefying wood using phenol with a phosphoric acid catalyst were used by Lin et al. [47] in the preparation of moldings by employing hexamine as a hardener. They investigated how the molding conditions and composition affected the flexural properties and water-sorption kinetics of the moldings. They reported that the liquefied wood resins, while composed of different kinds of wood components, exhibited satisfactory and nearly uniform curing reactivity. The flexural properties of the liquefied wood moldings were enhanced when the amount of combined phenol within the liquefied wood was increased. When the amount of combined phenol exceeded 75 %, these properties were comparable to those of commercial novolac. It was also observed that by increasing the wood fillers in the content, the flexural properties of the liquefied wood moldings were improved more effectively when compared with those seen in commercial novolac moldings. This demonstrated that a greater reinforcing effect than that of commercial novolac resins could be achieved for liquefied wood resins by compounding them with wood fillers. Moreover, the reinforcing performance of the wood fillers became greater as the combined phenol amount was increased. In addition, observations of the moldings, water-sorption measurements and scanning electron microscopy (SEM) revealed that the hydrophilicity of the liquefied wood resins was much greater than that of the novolac, thus indicating a greater compatibility with the wood fillers [47].

In order to study the value of using liquefied wood-based resin as a molding material Zhang et al. [48] liquefied Chinese fir (*Cunninghamia lanceolata*) and poplar (*Populus tomentosa*) wood meal in phenol. They co-condensed the reactant with formaldehyde to produce a liquefied wood-based resin. Their study examined the characterization of the resin and its application as a molding material. They found the basic properties of the liquefied wood-based resin to be within acceptable limits, and measured the bonding strength of plywood produced from the liquefied Chinese fir and poplar resins as 1.54 and 1.00 MPa, respectively. Their compression strengths of 73.01 and 73.58 MPa, respectively, were similar to that of phenol-formaldehyde (PF) resin molding material. The limiting volume swelling of the liquefied Chinese fir and poplar resin molding materials were 8.5 % and 8.3 %, the thickness swelling rates of water absorption 3.3 % and 4.2 %, and the maximum weight ratios of water absorption 25.9 % and 26.2 %, respectively. The soil burial test resulted in weight loss rates of 8.3 % and 9.1 % for the molding materials made from the liquefied Chinese fir and poplar resins, while that of the PF resin molding material was 7.9 %. After the soil internment test, the compression strength reduction ratios of the two kinds of molding material were seen as 16.9 % and 17.7 %, with that of the PF resin molding material at 15.4 %. The results of the wood fungi inoculation test on three surfaces of the molding materials indicated that the breeding rates on the liquefied Chinese fir and poplar resin molding materials were at Level 4 of the ISO standard, while the PF resin molding material was at Level 1 [48].

A two-stage procedure was employed to prepare a phenolated wood/phenol/formaldehyde co-condensed novolac-type resin by Lin et al. [49]. The wood was first liquefied with phenol using an acidic catalyst to prepare a phenolated wood. After liquefaction, an aqueous solution of formaldehyde (formalin) was introduced in order to carry out a co-condensation reaction to convert the remaining nonreacted phenol into resin components. With this procedure, they were able to convert almost all the phenol which remained after liquefaction into resin, thus significantly upgrading the practical value of their liquefaction method. In addition, this technique also demonstrated its capability of greatly enhancing the thermofluidity of phenolated wood resins and the mechanical properties of their molded products. The resulting co-condensed resin melt viscosities and flow temperatures were much lower than those of the phenolated wood resins. The similarity of these two properties to those of conventional novolac resin gives them excellent processability potential. The flexural properties were much higher in the co-condensed resin molded products than in the phenolated wood, and they were also somewhat better than those of conventional novolac resin, making this process a prospective technique for the preparation of wood-based novolac resins [49].

Pan [50] conducted a wood liquefaction study using phenol for the reagent solvent and in the presence of oxalic acid as the catalyst. A succession of studies has been carried out on liquefied wood, liquefied wood residues, and novolac-type liquefied wood resins (LWRs) and biocomposites produced from LWR. Results of LWR characterization have indicated that different liquefaction mechanisms are involved when liquefaction reactions are conducted in different types of reaction vessels. The crystallinity indices of LWRs were higher than that of the original wood, indicating that the wood component most vulnerable to the liquefaction reaction was the amorphous lignin. Ions of Fe^{2+} and Fe^{3+} were found to have catalytic effects during the liquefaction reaction. The cure kinetic investigation of two typical LWRs showed that the activation energies of LWR were higher than those of conventional phenolic resins, while being close to that of a lignin-phenol-formaldehyde resin from another study. It was concluded that LWR followed an autocatalytic cure mechanism and, based on isothermal DSC methods, two kinetic models were proposed for LWR. The flexural strengths of the composites were comparable to those of similar products reported by other researchers, indicating that the LWR and liquefied wood residue produced via weak acid catalyzed liquefaction could be successfully applied to molded biocomposite products in place of conventional novolac resin [50].

The PF resin manufacturing industry has been facing growing health concerns over the use of the carcinogen formaldehyde, as well as being faced with sustainability issues dealing with the use of petroleum-based phenol to produce PF resin. In the synthesis of phenolic novolac resins, glucose and the glucose derivative 5-hydroxymethylfurfural (5-HMF) have been demonstrated to be potential substitutes for formaldehyde. A number of studies have examined glucose and 5-HMF resin systems including a phenol-glucose (PG) novolac resin curing process using a bisphenol-A type epoxy. The modeling of the process was based on the Sestak–Berggren equation (SB) employing Málek methods, followed by a novel one-pot, two-step method for synthesizing phenol-5-hydroxymethylfurfural (PHMF) resin using phenol reacted with in situ generated HMF from glucose [51].

Research on the catalytic effect of $CrCl_2/CrCl_3$ and tetraethylammonium chloride (TEAC) in the process found that it facilitated glucose dehydration as well as phenol-aldehyde polycondensation reactions. The average molecular weight (M_w) of these PHMF resins was in the range of 700–900 g/mol and their structure was similar to novolac PF resins. Moreover, an attempt was made to use lignin as a bio-based curing agent for the synthesized PHMF resin in order to apply a greener approach to the system. The curing mechanism was examined using spectral techniques and a lignin model compound. Upon curing, a void-free polymer matrix was prepared from the PHMF and bis-phenol-A epoxy resins. In addition, bio-phenolic compounds have been produced from woody biomass via hydrothermal liquefaction or hydrolytic depolymerization of lignin. In the production of bio-based PHMF (BPHMF) resins, these biophenolic feedstocks have been applied to partially substitute for the petroleum-based phenol. Differential scanning calorimetry (DSC) has been useful in studying the curing kinetics of all the crosslinking reactions of PHMF and BPHMF resins [51].

The results of model-free methods and model-predicted reaction rates were in good agreement with the experimental results for the kinetic parameters. The optimum amount of hardener (i.e., curing agent) was determined by reviewing the effects of different process parameters including curing temperature and reaction time. Glass fiber reinforced PHMF and BPHMF resin composites were then produced by impregnating glass fibers with the PHMF or BPHMF resins, and their thermal, mechanical, dynamic mechanical and rheological properties were examined along with the chemical and water resistance of the matrix and the composites [51].

The rapid liquefaction of wood (over 90 %) at the supercritical temperature of phenol for 0.5 min was achieved by Lee and Ohkita [52]. The combined phenol content of this liquefied wood was about 75 %. The effects of a variety of reaction conditions on liquefaction were studied. As the reaction temperature, the phenol/wood weight ratio, and the charged mass-to-reactor capacity (w/v) ratio increased, the amount of methanol-insoluble residue decreased, whereas the combined phenol content increased. After sufficient liquefaction had been achieved, the molecular weights and the polydispersity of the products obtained were 400–600 and 1.5–2.5, respectively. Early in the reaction, the molecular weight of the wood components was markedly lower, but increased slightly as reaction time increased. The properties of the liquefied wood were examined and compared with those resulting from conventional liquefaction methods. The combined phenol content was observed to be similar to that obtained by other liquefaction methods, with the exception of the sulfuric acid-catalyzed method. This technique, however, produced flow properties comparable to those of other liquefaction methods. The flexural strength of liquefied wood moldings was also equivalent to that of moldings prepared using other liquefaction methods [52].

A few studies were carried out about phenolic resins synthesized in the presence of alkaline catalysts. Alma et al. [11] investigated the catalytic effects of various alkaline and metallic salts on liquefaction of biomass into phenol. They reported that most alkalis and metallic salts were efficient catalysts giving low biomass residue content, however, they were not effective catalysts to achieve a high amount of combined phenol. NaOH was found to be the most effective base catalyst in the liquefaction of biomass in phenol at an elevated temperature of 250 °C.

In another study Alma et al. [40] prepared NaOH-catalyzed phenolated wood-phenol-formaldehyde co-condensed novolac-type resin in two stages. Birch wood meal was first phenolated by using NaOH as a catalyst an elevated temperature (i.e., 250 °C) and then 37 % formaldehyde solution together with oxalic acid was added to the phenolated wood-phenol mixture obtained after the phenolation. The yields obtained due to resinification of phenolated wood were found to depend, greatly, upon phenol-formaldehyde molar ratio. Moreover, the melt flow properties (i.e., viscosity) of the resinified NaOH-catalyzed phenolated wood reached or slightly exceeded those of commercial novolac resin.

1.4.2 Epoxy-type resins from liquefied biomass

Asano et al. [53] prepared liquefied products with high wood content by pretreating wood with ozone before liquefaction. The findings showed that it was possible to increase the wood content in the wood/polyhydric alcohol solvent ratio of 2 : 1. They produced resin blends by mixing liquefied products with water-soluble ethylene glycol diglycidyl ether (EGDGE) and oily consistency diglycidyl ether of bisphenol A (DGEBA). It was possible to increase the wood content of the resin blend to 53 %. The resins were cured by citric acid or triethylene tetramine (TETA), and their mechanical properties were evaluated. Dynamic mechanical measurements revealed that the former had higher glass transition temperatures than the latter. It was found that the resin with DGEBA cured by citric acid had nearly the same level of tensile strength as commercial plastics.

Wu and Lee [54] reacted liquefied bamboos with bisphenol A and epichlorohydrin and utilized the obtained product to prepare copolymer epoxy resins. They investigated the curing and thermal properties of obtained resins. According to the results the copolymer epoxy resins could cure at room temperature after the hardener was added, and their curing process was exothermic. During curing, the copolymer epoxy resins obtained from phenol-liquefied bamboo had higher heat released than those obtained from polyhydric alcohol-liquefied bamboo. The DSC analysis showed that heat treatment could enhance the crosslinking of copolymer epoxy resins cured at room temperature. However, resins prepared with polyhydric alcohol-liquefied bamboo had a lower glass transition temperature. The TGA analysis showed that resins prepared with phenol-liquefied bamboo had better thermal stability.

Zhang et al. [55] prepared liquefied corn barn-based epoxy resin (LCBER) through the glycidyl etherification reaction from liquefied corn barn (LCB), which has groups of bound phenol, and epichlorohydrin under alkali conditions. They examined the average molecular weights of LCB and LCBER in various liquefaction conditions. They found that the macromolecular chain and epoxy function of the resins would be a dominant factor for crosslinking density and properties of the cured LCBER. The cured liquefied CB-based epoxy resin (LCBER-30) using the corresponding LCB at 30 min (LCB-30) exhibited higher glass-transition and showed higher decomposition temperatures at 5 % weight loss compared with the raw materials having much more macromolecular structure. However, its shear strength was lower in comparison with the other LCBER ones. Fig. 1.8 shows the general stages, methods and chemicals used for the production of liquefied biomass based epoxy resin.

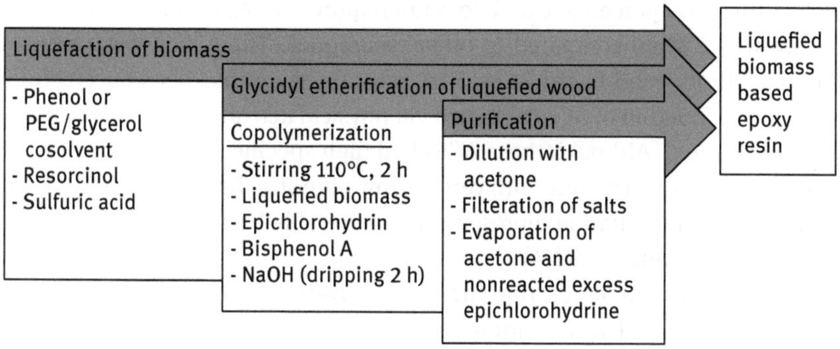

Fig. 1.8: Epoxy-type resin production procedures from liquefied biomass.

1.5 Fibers

Activated carbon fibers (ACFs) have superb features due to their developed pore structure, large specific surface area, pore volume, and uniform microporosity and special surface reactivity compared with the common activated carbons. These materials have been greatly used in many areas such as adsorption and separation, catalyst support, and electronic materials [56]. ACFs have mainly been produced via carbonization of various fiber precursors, such as rayon, petroleum or coal-based pitch, polyacrylonitrile (PAN) and commercial phenol formaldehyde resin, etc. [57]. However, limited reserves of these resources, the costs of processing and raw material supply, yield, environmental issues, etc. make it necessary to improve the production of ACFs that highly depend on chemical materials [58]. Therefore, to solve these problems, utilization of biomass resources became an important hot issue and many techniques are being developed currently in order to use biomass resources effectively.

To expand the usage area of liquefied wood and obtain highly value-added products, some scientists studied carbon fibers and carbon fiber precursors from liquefied wood [59]. With the exception of wood, some biomass-based ACFs were also produced from renewable sources, such as bamboo, kenaf, jute, cotton stalk, etc. become popular as an alternative to petrochemical resources [24]. On the other hand, in some of the studies, cellulose [60] and lignin [61] have been used as precursors for ACF production. Cellulose derived ACFs were first developed in the 1950s and 1960s. Then, cellulose became the third largest resource for ACFs after polyacrylonitrile (PAN) and petroleum pitch. However, demand for cellulose-derived ACFs reduced because of their low strength and low yield for high-resistance applications compared with PAN and pitch-based carbon fibers [62].

On the other hand, wood, contains plenty of cellulose and lignin, has been phenolated for the production of ACFs via the liquefaction method. This procedure gives a higher wood utilization percentage and does not require the separation of main components from wood when compared to other techniques. There have been several studies conducted on wood-based ACFs. Very recently, wood has been successfully liquefied for the production of ACFs obtained via physical activation method by using steam [63] and CO_2 [64]. Although these ACFs had high specific surface area, most of them have a microporous structure with very narrow pore size distribution (PSDs) and they had low yield. This situation represents a considerable barrier for their usage as catalyst supports and capacitors [24].

Therefore, some scientists used a chemical activation method, because it has some advantages over physical activation such as defined micropore size distribution, high yield, pore volume and specific surface area due to reaction with anisotropic carbons to improve large porosity [65, 66]. Some authors prepared jute based ACFs with a porosity of 76.14 by phosphoric acid activation and they investigated the removal of methylene blue (MB) from aqueous solution. ACFs were successfully employed as an adsorbent for the quantitative removal of MB from aqueous solution [67]. The general steps of CF or ACF production by using liquefied biomass are illustrated in Fig. 1.9.

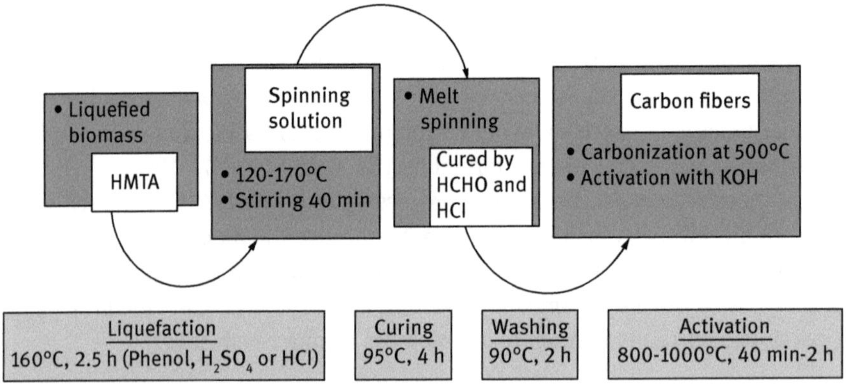

Fig. 1.9: General procedures of ACF production from liquefied biomass.

1.5.1 Fibers from liquefied lignocellulosics

Ma and Zhao [68] created liquefied wood in phenol fibers (WPFs) by spinning liquefied wood spinning solution (LWS) including hexamethylenetetramine (HMTA) as a hardener. The fibers thus obtained were then cured by soaking in a solution containing the chief components of hydrochloric acid and formaldehyde. The chemical structure of the WPFs, as identified by FT-IR spectrometry, was extremely changed from that of the liquefied wood (Fig. 1.10). The WPFs, having an average diameter of ≈ 27–42 μm, tensile strength of ≈ 230–356 MPa, and modulus of ≈ 15–31 GPa, were obtained using a spinning speed of 0.72 μm min^{-1}, hydrochloric acid concentration of 18.5 %, heating rate of 10 °C h^{-1}, and curing time of 4 h. The high thermal stability and complex thermal decomposition process of these WPFs was confirmed by TGA. In addition, two noticeable weight loss temperatures of WPFs were observed as 510 °C and 748 °C.

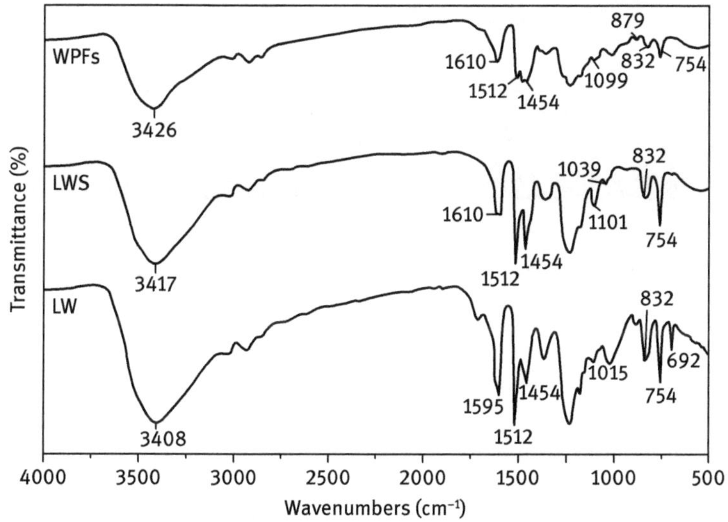

Fig. 1.10: FT-IR spectra of LW, LWS, and WPFs [68].

1.5.2 Carbon and activated-carbon fiber from liquefied lignocellulosics

Activated-carbon fibers from liquefied wood were produced via a steam activation process. Through the use of burning behavior analysis, X-ray diffraction, nitrogen adsorption-desorption isotherms, X-ray photoelectron spectroscopy, and SEM, researchers studied the ways in which activation temperature and time affected the microstructure and surface functional groups of the liquefied wood activated-carbon fibers (LWACFs). The results showed that with an increase in temperature or time, there was a corresponding gradual increase in the burn-off value of the LWACFs. None

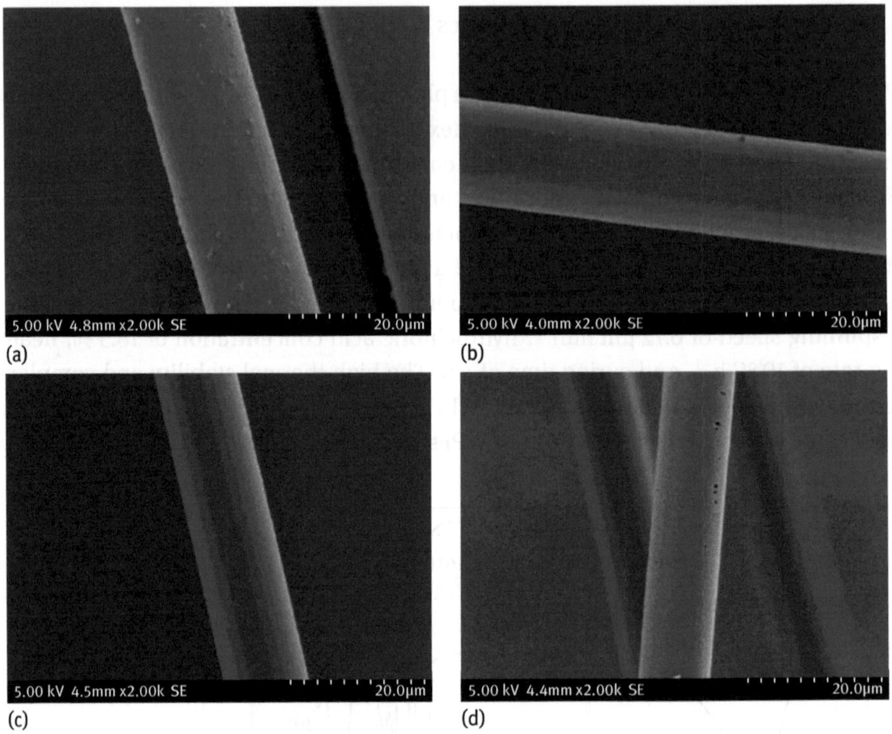

Fig. 1.11: SEM images of the LWACF samples with different temperatures (a) 700–60, (b) 750–60, (c) 800–60, (d) 850–60 [63].

of the LWACF structures were close to being graphitized, and generally, the amount of graphitization and the thickness of the crystal structure increased when there was an increase in temperature or time [63]. Moreover, the LWACFs contained many micropores, their specific surface area, pore volume, micropore size, and mesopore quantity being directly correlated with the activation time or temperature (Fig. 1.11).

The greatest specific surface area was 2641 m^2/g. In all the samples, the fractal dimension values of nearly 3 indicated that their surfaces were quite rough. Additionally, the elemental carbon content increased with an increase in temperature or time, while that of oxygen decreased. The relative graphitic carbon content also decreased as the temperature or time increased, whereas that of carbon bonded to oxygen-containing functions increased. A great number of holes were formed in the surface of samples prepared at a higher temperature or over a longer time [63].

Douglas fir (*Pseudotsuga menziesii*) wood flour was liquefied using phenol/ catalytic amounts of sulfuric acid mixture, after which formalin was added and reacted. In order to obtain the liquefied resinified wood, the free phenol was distilled off from the reaction mixture under reduced pressure and the resulting wood was then melt spun, stabilized via three-dimensional curing and finally carbonized at 900 °C. Carbon fiber was produced from liquefied wood without reacting with formaldehyde (Fig. 1.12). The resulting formation of macropores in the fiber was observed under SEM (Fig. 1.13). However, when adequate amounts of formaldehyde were employed in the resinification of the liquefied wood, the formation of the macropores was prevented. This was validated by the carbonization of synthetic novolac fibers. Moreover, the thermogravimetric analysis (TGA) measurements of the carbonization yield also agreed with these observations [69].

Xiaojun and Guangjie [23] prepared carbon fibers from liquefied wood by adding hexamethylenetetramine and soaking them in a solution chiefly consisting of hydrochloric acid and formaldehyde. The structural transformation of carbon fibers from LWCFs was examined via Fourier transform infrared (FT-IR) spectroscopy and X-ray powder diffraction (XRD). The results showed that after carbonization, the structure of the precursor fibers from the liquefied wood had been completely changed. The apparent crystallite size of Lc (002) and layerplane length parallel to the fiber axis of La (100) gradually increased during carbonization. Certain carbonization conditions resulted in carbon fibers with a maximum tensile strength of 1.7 GPa and at the same time enabled the determination of the critical stage at which the specific surface area of LWCFs changes as 600–800 °C.

1.6 Films and coatings

1.6.1 Liquefaction of biomass for polyester production

Liquefied biomass can be considered a polyhydric alcohol. Kunaver et al. [70] reported that lignocellulosics including medium-density fiberboard, veneered particleboard, particleboard, oriented strand board, plywood, spruce (*Picea* spp.), and wheat straw powders had been liquefied by using an ultrasound-assisted method, a diethylene glycol/glycerol mixture (1 : 4 w/w) and p-toluenesulfonic under atmospheric pressure. In the ultrasound-assisted method, the high-frequency (24 kHz) power output was regulated by adjusting the amplitude from 20 % to 100 % of the nominal power of 400 W. The high-frequency output was transferred through a titanium cylindrical probe inserted into the reactor through the side neck and immersed 20 mm into the reaction mixture. The mixture was heated at 180 °C for 2 h under stirring and upon the reaction mixture reaching 160 °C, the ultrasound was switched on. Liquefied wood was employed as a component in polyester synthesis because of its large number of available hydroxyl groups (500–1000 mg of KOH/g) [70].

The liquefied wood was used as a replacement for part of the polyhydroxy alcohols, which are the standard raw material in polyester formulation. Liquefied wood-based polyesters were prepared with adipic acid and/or phthalic acid anhydride at the high temperature of 200 °C via a polycondensation/esterification reaction using dibutyl tin oxide (0.2 % w/w) as the catalyst for 160–180 min under stirring. A slight

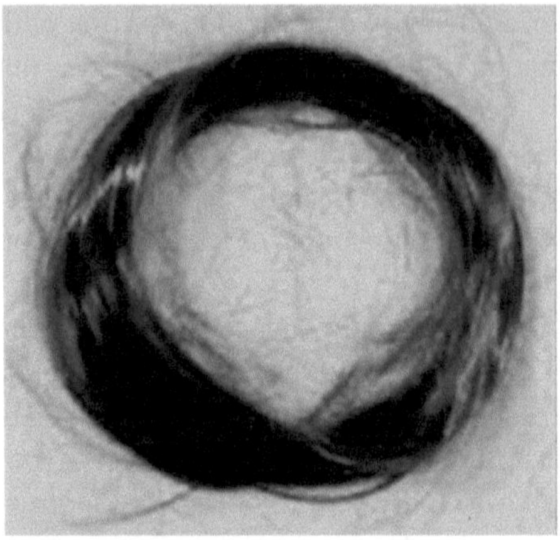

Fig. 1.12: Visual appearance of the melt spun fibers [69].

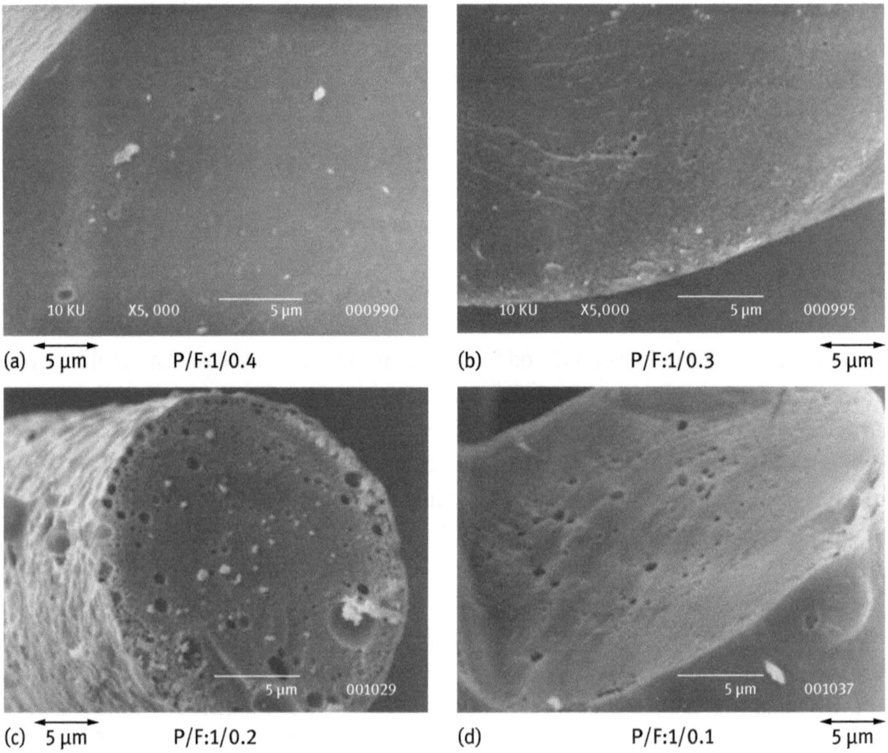

Fig. 1.13: SEM images of the fracture cross sections of carbonized fibers with different phenol/formaldehyde molar ratio [69].

stream of nitrogen was introduced into the reactor to enable easier transport of water vapor into the condenser. A modification of the polyester reactivity and complexity such as this shows promise for its further utilization in polyurethane synthesis. The final OH values of the polyesters were in the same range as the saturated polyesters used in polyurethane film and coating production. The viscosity greatly depends on the type of glycol used for liquefaction and on the wood/glycol ratio. In these experiments, biomass was used in place of 23 % of the polyhydroxy alcohols in the polyester. Some benefits should be expected from systems like this which incorporate biomass components into polymeric compositions and, consequently, provide a certain degree of biodegradability. Esterification or other self-condensation of a proportion of the hydroxyl groups lessens the reactivity of the liquefied wood. The results revealed that the liquefied wood-based polyesters had viscosity properties similar to those possessed by equivalent commercial polyesters [70].

Spruce wood meal was liquefied using a mixture of diethylene glycol and glycerol with a minimal addition of p-toluenesulfonic acid as the catalyst. The liquefied wood was used as a replacement for a certain amount of the polyhydroxy alcohol in the synthesis of polyester. The replacement was enabled by the large number of hydroxyl groups available in the liquefied wood. Three different polyesters were synthesized by using adipic acid and phthalic acid anhydride as reagents and dibutyl tin oxide as the catalyst. Due to esterification, the hydroxyl values of the polyesters were reduced from 1043 mg KOH/g of the liquefied wood to 400–800 mg KOH/g. Wood derivatives were used to replace the polyhydroxyl alcohols (22–23 %) in the polyester formulations. These saturated polyesters would be suitable in polyurethane foam production applications [71]. Fig. 1.14 demonstrates the main procedures and reagents for the production of liquefied biomass based polyester.

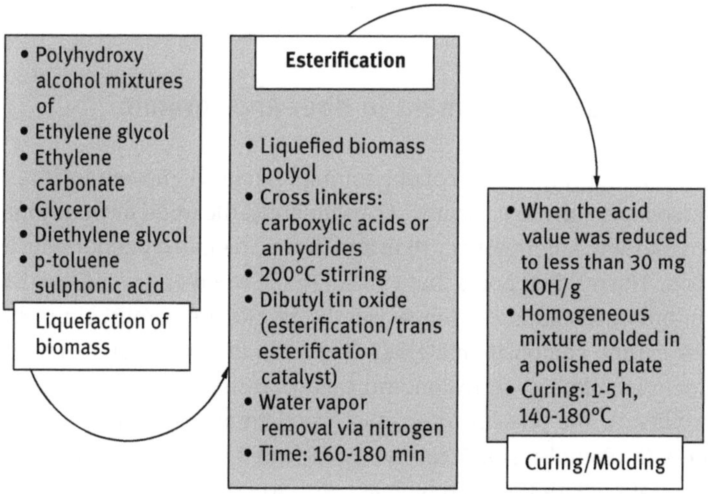

Fig. 1.14: The main steps of polyester production from liquefied biomass.

1.6.2 Liquefied biomass as replacement for polyurethane films

Cheumani-Yona et al. [72] prepared and characterized aesthetically pleasing light-colored polyurethane wood coatings produced from liquefied wood. Liquefied black poplar (*Populus nigra*) wood waste was obtained via solvolysis in a polyethylene glycol/glycerol mixture and then bleached with hydrogen peroxide. The liquefied wood was converted from a dark brown to a yellowish-colored product by the bleaching treatment. The liquefied wood was cured with polymeric diphenylmethane diisocyanate or trimethylolpropane toluene diisocyanate (TMP/TDI) in order to prepare polyurethane films. The mechanical properties of the films prepared from bleached liquefied wood with the TMP/TDI curing agent were comparable to films of unbleached liquefied wood and thus were deemed suitable for film applications. The addition of n-octyltriethoxysilane improved the film's poor resistance to water and ethanol and enhanced its hardness.

1.6.3 Self-crosslinking film from liquefied biomass

Budija et al. [73] used diethylene glycol (DEG) to liquefy black poplar wood with 3 % sulfuric acid as the catalyst at 150 °C for 95 min at a wood/DEG ratio of 1 : 5. The unreacted DEG was effectively separated from the excess solvent-free liquefied wood (ESFLW) and its levulinic acid content analyzed via HPLC. The hydroxyl number indicated that the ESFLW in the liquid mixture was a factor in up to 60 % of the free single-bond OH groups. The crosslinking of ESFLW without the use of curing agents or additives was performed for the first time, and the drying stages were then examined. Analyses using FT-IR spectroscopy demonstrated that this crosslinked polymer film could be classified as an ether and/or ester network.

1.7 Liquefied wood as replacement in fiber and ceramic

Wood ceramics are commonly fabricated of phenol-impregnated lignocellulosic resins carbonized at high temperature in a vacuum. Consequently, the wood or woody material is transformed into soft, amorphous carbon and the impregnated phenol resin into hard, glassy carbon. The original pores that existed in the wood and woody material are still present in the wood ceramics. Hence, wood ceramics are porous, amorphous carbon and glassy carbon composite materials (Fig. 1.15). Positive characteristics of wood ceramics include stiffness, corrosion and friction resistance, and electromagnetic shielding ability. At the present time, most wood ceramics are manufactured via impregnation with phenol resin. A review investigated the production of wood ceramics by impregnation with liquefied biomass with wood powder, resulting in replacement of phenolic resin with liquefied wood resin [74].

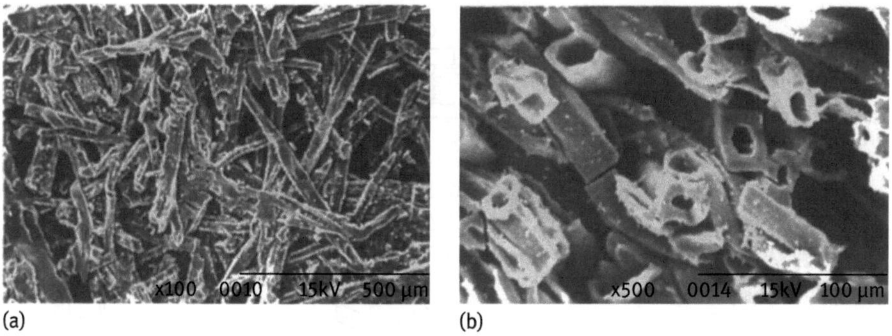

Fig. 1.15: SEM pictures of wood ceramics at 2500 °C. (a) surface, (b) cross section [74].

Hirose et al. [75] phenolated hinoki (*Chamaecyparis obtusa* Endl.) wood powder at 150 °C using sulfuric acid as the catalyst. Medium-density fiberboard (MDF) samples 20 × 20 × 15 mm in size and having an original density of 0.58 g/cm^3 were impregnated with liquefied wood diluted with ethanol (ratio 1 : 1) under vacuum atmosphere. The MDF was then carbonized in a vacuum furnace to make wood ceramics. The material being processed generated a considerable amount of decomposition products (e.g., pyroligneous acid, wood tar, etc.). In order to ensure the quality of the wood ceramics, the carbonization was carried out in a vacuum furnace where the decomposition product was vented off. This study used the carbonization temperatures of 400 °C, 500 °C, 650 °C, and 800 °C to develop a new wood ceramics product impregnated with liquefied wood and then investigated how carbonization temperature affected its dimensional shrinkage, weight loss, density, compressive strength and volume electrical resistivity. The results showed that with the rise in carbonization temperature, the dimensional shrinkage increased, and as compressive strength increased, electrical resistivity decreased [75].

Hirose et al. [76] liquefied hinoki (*Chamaecyparis obtusa* Endl.) wood powder to phenol by using an acidic catalyst (sulfuric acid) at 150 °C. Soft carbon fiber was produced by carbonizing bamboo pulp fiber at 800 °C for 2 h. This carbonized bamboo fiber was then mixed with the liquefied wood diluted with ethanol at a ratio of 2 : 1 (Fig. 1.16). Next, it was dried and formed into board at 160 °C under pressure of 5 MPa for 10 min. The ratio of carbonized bamboo fiber to liquefied wood was 1 : 1.3. The board was carbonized in a vacuum furnace to form wood ceramics. During carbonization, the material generated a considerable amount of decomposition products including pyroligneous acid and wood tar; therefore, in order to ensure the quality of the wood ceramics, the carbonization was carried out in a vacuum furnace where the decomposition products could be vented off continuously. The carbonization temperatures used were 400 °C, 500 °C, 650 °C, and 800 °C. The authors of this study developed a new wood ceramics product from carbonized bamboo fiber and liquefied wood, and investigated the effects of carbonization temperature on the dimensional

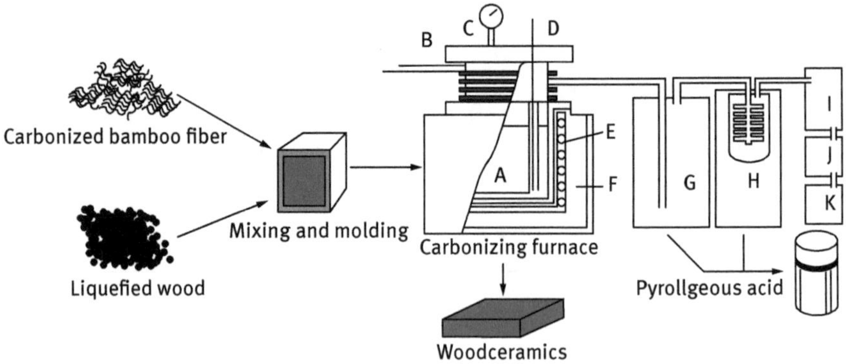

A Material	B Cooling tube	C Pressure gauge	D Thermometer
E Heater	F Finsutation material	G Water trap	H Cold trap
I Activated carbon filter	J Air filter	K Vacuum pump	

Fig. 1.16: Wood ceramics manufacturing methods [76].

shrinkage, weight loss, density, compressive strength, and volume electrical resistivity of the wood ceramics. The results showed that liquefied wood decomposed with increasing carbonization temperature and then combined with the carbonized bamboo fiber. The dimensions of the samples remained nearly constant, the compressive strength was improved and electrical resistivity was reduced [76].

Zhao et al. [77] produced composites using wood powder and its phenolated product which were subsequently carbonized into wood ceramics. They examined how the composite contents in the phenolated wood affected the forming ability, size, density, compressive strength, volume electrical resistivity, and specific surface area of the wood ceramics. Three parts of phenol were mixed with 0.04 part 98 % sulfuric acid and one part wood powder. The mixture was stirred at 150 °C for 3 h to produce a black liquid- liquefied wood. This was then washed several times using distilled water to remove the free phenol. Finally, purple phenolated wood was prepared. The phenolated wood was milled into powder less than 63 μm in size, and was then further ball-milled into powder with an average size of about 10 μm. This phenolated wood powder was mixed with the wood powder in different proportions and mold-pressed 5.5 g of each mixture at room temperature under pressure of 10.0 MPa for 5 min into cylindroid composites having the uniform size of 20.0 mm × 20.0 mm. The wood ceramics were then produced by carbonization of the cylindroid composites in a vacuum furnace in which the temperature was raised at a rate of 1 °C/min, and then maintained for 2 h at 800 °C. The proportions by weight of phenolated wood powder to wood powder were 100 : 5, 100 : 10, 100 : 20, 100 : 30, 100 : 50, 100 : 70, 100 : 100, 70 : 100, 50 : 100, 30 : 100, 20 : 100, and 10 : 100, respectively. It was suggested that in the preparation of wood ceramics, phenolated wood, as the main constituent of liquefied wood, can possibly be used to totally replace the phenolic resin [77].

1.8 Conclusion

Today, the majority of fossil fuels such as oil, coal and natural gas are located within the boundaries of a few countries and very little will remain of their reserves in the near future. In addition, petroleum-derived plastics contribute to environmental issues, particularly for aquatic life because, being non-biodegradable, they may continue to contaminate the ecosystem for hundreds of years. However, biomass is a cleaner alternative to conventional fossil fuels because it contains only insignificant amounts of sulfur, nitrogen and ash and thus produces lower emissions of SO_2, NOx and smoke. At present, mechanical, thermochemical, biochemical and chemical techniques can be applied for the conversion of biomass biofuels as well as useful green chemicals and materials. Nevertheless, in order to achieve sustainable and renewable production, it is essential that scientists focus on investigating new liquefaction approaches and techniques for improving the cost, yield and duration of the process and the range and features of the products.

There are several approaches in the utilization of the biomass wastes such as production of energy, oil, chemicals, plastics, and composite materials. The production of composite materials which consist of chemical modification or liquefaction of biomass wastes appears to be one of the ideal approaches in the utilization or re-use of the wastes, in terms of environment, economy, and availability of raw material. In order to achieve a more complete utilization of biomass wastes, a variety of studies are being intensively promoted on wood chemicals, modified natural polymers, energy, and new pulping methods. Apart from those, other novel methods on the utilization of biomass wastes are plasticization by conventional chemical processing such as etherification, esterification and so on, and solvolysis of wood, so-called “liquefaction of wood”, for making wood waste-based plastic sheets, molded products, foams, adhesives etc.

The utilization of biomass wastes, which are the most abundant natural source, and renewable and cheap materials, is becoming an important issue for many researchers nowadays. Of many trials, the liquefaction of biomass wastes seems to be one of the effective ways to utilize wastes due to its applicability to various kinds of materials. A significant majority of the main transportation fuels and new type adhesives, foams and molded wood products could be obtained from liquefied biomass, particularly biomass wastes, in the near future. As a result, biomass liquefaction has great potential, for this reason new and renewable approaches and techniques have to be developed to improve the efficiency of the process and the properties of the products.

References

[1] Stöcker M. Biofuels and biomass-to-liquid fuels in the biorefinery: Catalytic conversion of lignocellulosic biomass using porous materials. Angew Chem Int Ed. 2008;47:9200–9211.
[2] Serrano-Ruiz JC, Dumesic JA. Catalytic routes for the conversion of biomass into liquid hydrocarbon transportation fuels. Energy Environ Sci. 2011;4:83–99.
[3] Mapping the Development of UK Biorefinery Complexes: Literature Review, 2007. (Accessed May 2, 2007, at http://www.nnfcc.co.uk/tools/mapping-the-development-of-uk-biorefinery-complexes-literature-review-nnfcc-07-008-)
[4] Xiu S, Zhang B, Shahbazi A. Biorefinery processes for biomass conversion to liquid fuel. In: Dos Santos MAB, editor. Biofuel's engineering process technology. Rijeka, Croatia: InTech; 2011. p. 167–190.
[5] Demirbas A. Biorefineries: Current activities and future developments. Energ Convers Manage. 2009;50:2782–2801.
[6] Demirbas A. Mechanisms of liquefaction and pyrolysis reactions of biomass. Energ Convers Manage. 2000;41:633–646.
[7] Toor SS, Rosendahl L, Rudolf A. Hydrothermal liquefaction of biomass: A review of subcritical water technologies. Energy. 2011;36:2328–2342.
[8] Pan H. Synthesis of polymers from organic solvent liquefied biomass: A review. Renew Sustain Energ Rev. 2011;15:3454–3463.
[9] Bensaid S, Conti R, Fino D. Direct liquefaction of ligno-cellulosic residues for liquid fuel production. Fuel. 2012;94:324–332.
[10] Alma MH, Yoshioka M, Yao Y, Shiraishi N. Preparation of oxalic acid catalyzed resinified phenolated wood and its characterization. Mokuzai Gakkaishi. 1995;41:1122–1131.
[11] Alma MH, Yoshioka M, Yao Y, Shiraishi N. Preparation of sulfuric acid-catalyzed phenolated wood resin. Wood Sci Technol. 1998;32:297–308.
[12] Alma MH, Basturk MA. Co-condensation of NaOH catalyzed liquefied wood wastes, phenol, and formaldehyde for production of resol type adhesives. Ind Eng Chem Res. 2001;40:5036–5039.
[13] Nui M, Zhao G, Alma MH. Thermogravimetric studies on condensed wood residues in polyhydric alcohols liquefaction. BioRes. 2001;5:615–630.
[14] Alma MH. Several acids-catalyzed phenolation of wood and its application to molding materials. PhD thesis, Kyoto University, Kyoto, Japan. 1996.
[15] Yu F, Liu Y, Pan X, Lin X, Liu C, Chen P, Ruan Y. Liquefaction of corn stover and preparation of polyester from the liquefied polyol. Appl Biochem Biotechnol. 2006;129:574–585.
[16] Liang L, Mao Z, Li Y, Wan C, Wang T, Zhang L, Zhang L. Liquefaction of crop residues for polyol production. BioRes. 2006;1:248–256.
[17] Acemioglu B, Alma MH. Kinetics of wood phenolysis in the presence of HCl as catalyst. J Appl Polym Sci. 2002;85:1098–1103.
[18] Alma MH, Basturk MA. Liquefaction of grapevine cane (*Vitisvinisera L.)* waste and its application to phenol-formaldehyde type adhesive. Ind Crop Product. 2006;24:171–176.
[19] Alma MH, Yoshioka M, Yao Y, Shiraishi N. The preparation and flow properties of HCl catalyzed phenolated wood and its blends with commercial novolak resin. Holzforschung. 1996;50:85–90.
[20] Alma MH, Shiraishi N. Preparation of polyurethane-like foams from NaOH catalyzed liquefied wood. Holz Roh Werkstoff. 1998;56:245–246.
[21] Lee SH, Teramoto Y, Shiraishi N. Resol-type phenolic resin from liquefied phenolated wood and its application to phenolic foam. J Appl Polym Sci. 2002;84:468–472.
[22] Xie T, Chen F. Fast liquefaction of bagasse in ethylene carbonate and preparation of epoxy resin from the liquefied product. J Appl Polym Sci. 2005;98:1961–1968.

[23] Xiaojun M, Guangjie Z. Preparation of carbon fibers from liquefied wood. Wood Sci Technol. 2010;44:3–11.
[24] Ma X, Zhang F, Zhu J, Yu L, Liu X. Preparation of highly developed mesoporous activated carbon fiber from liquefied wood using wood charcoal as additive and its adsorption of methylene blue from solution. Bioresour Technol. 2014;164:1–6.
[25] Pilato L. Phenolic resins: A century of progress. Berlin, Heidelberg, USA: Springer-Verlag; 2010.
[26] Alma MH, Basturk MA. New polyurethane-type rigid foams from liquified wood powders. J Mater Sci Lett. 2003;22:1225–1228.
[27] Gao LL, Liu YH, Lei H, Peng H, Ruan R. Preparation of semirigid polyurethane foam with liquefied bamboo residues. J Appl Polym Sci. 2010;116:1694–1699.
[28] Zhang JP, Du MH, Hu LS. Bamboo liquefaction with polyhydric alcohols and its application in flexible polyurethane foam. Adv Mater Res. 2012;524:2113–2117.
[29] Yaoguang Y, Yoshioka M, Shiraishi N. Water absorbing polyurethane foams from liquefied starch, J Appl Polym Sci. 1996;60:1939–1949.
[30] Hakim AAA, Nassar M, Emam A, Sultan M. Preparation and characterization of rigid polyurethane foam prepared from sugar-cane bagasse polyol. Mater Chem Phys. 2011;129:301–307.
[31] Hu S, Wan C, Li Y. Production and characterization of biopolyols and polyurethane foams from crude glycerol based liquefaction of soybean straw. Bioresour Technol. 2012;103:227–233.
[32] Lee SH, Teramoto Y, Shiraishi N. Biodegradable polyurethane foam from liquefied waste paper and its thermal stability, biodegradability, and genotoxicity. J Appl Polym Sci. 2002;83:1482–9.
[33] Ge JJ, Wu R, Deng BL, Shi XH, Ming W, Li WJ. Studies on the biodegradable polyurethane materials based on bagasse (I) The liquefaction of bagasse and preparation of polyether ester polyol. Polym Mater Sci Eng. 2003;19:194–198.
[34] Ertaş M, Fidan MS, Alma MH. Preparation and characterization of biodegradable rigid polyurethane foams from the liquefied eucalyptus and pine woods. Wood Res. 2014;59:97–108.
[35] Huang Y, Zheng Z, Feng H, Pan H. Phenolic foam from liquefied products of walnut shell in phenol. Adv Mater Res. 2011;236–238:241–246.
[36] Alma MH, Shiraishi N. Preparation of sulfuric acid-catalyzed phenolated wood resin. J Polym Eng. 1997;18:179–196.
[37] Mcelhinney TR. Synthetic resin and process of making same. US Patent Number: US2394000 A; 1946.
[38] Hesse W, Jung A. Hardenable binding agents and their use. EPO Patent 0043097-A1; 1982.
[39] Alma MH. Determination of the biodegradation of phenolated wood-based molding materials by strength loss method. Holz Roh Werkst. 2008;66:237–239.
[40] Alma MH, Maldas D, Shiraishi N. Resinification of NaOH-catalyzed phenolated wood–phenol mixture with formalin for making molding materials. J Adhes Sci Technol. 2002;16:1141–1151.
[41] Shiraishi N. Plasticization of wood and its application. In: Shiraishi N, Kajita H, Norimoto M, editors. Recent research on wood and wood-based materials. London and New York: Elsevier Applied Science; 1993. p. 155–167.
[42] Candan Z, Gonultas O, Akbulut T, Balaban Ucar M. Development of carbon nanotube reinforced biocomposites for aerospace and automotive applications. 67th Forest Products Society International Convention, TX, USA; 2013.
[43] Candan Z, Gonultas O, Akbulut T, Balaban Ucar M. Characterization of carbon nanotubes reinforced biocomposites. 3rd International Non-wood Forest Products Symposium, Kahramanmaras, Turkey; 2014.

[44] Alma MH, Kelley SS. Conversion of the barks of several tree species into bakelite-like thermosetting materials by their phenolysis. J Polym Eng. 2000; 20:365–379.
[45] Alma MH, Kelley SS. Thermal stability of novolak type resins made by condensation of bark and phenol. Polym Degrad Stab J. 2000; 68:413–418.
[46] Alma MH, Kelley SS. Tensile properties of molding materials obtained by the condensation of various tree barks and phenol by using sulfuric acid as a catalyst. J Mater Sci. 2000;19:1517–1520.
[47] Lin L, Yoshioka M, Yao Y, Shiraishi N. Physical properties of moldings from liquefied wood resins. J Appl Polym Sci. 1995;55:1563–1571.
[48] Zhang QH, Zhao GJ, Yu LL, Jie SJ. Preparation of liquefied wood-based resins and their application in molding material. Forestry Studies in China. 2007;9:51–56.
[49] Lin L, Yoshioka M, Yao Y, Shiraishi N. Preparation and properties of phenolated wood/phenol/formaldehyde cocondensed resin. J Appl Polym Sci. 1995;58:1297–1304.
[50] Pan H. Wood liquefaction in the presence of phenol with a weak acid catalyst and its potential for novolac type wood adhesives. PhD thesis, Graduate Faculty of the Louisiana State University and Agricultural and Mechanical College, USA. 2007.
[51] Zhang Y. Production and applications of formaldehyde-free phenolic resins using 5-hydroxymethylfurfural derived from glucose in-situ. PhD thesis, The School of Graduate and Postdoctoral Studies, The University of Western Ontario. 2015.
[52] Lee SH, Ohkita T. Rapid wood liquefaction by supercritical phenol. Wood Sci Technol. 2003;37:29–38.
[53] Asano T, Kobayashi M, Tomita B, Kajiyama M. Syntheses and properties of liquefied products of ozone treated wood/epoxy resins having high wood contents. Holzforschung. 2007;61:14–18.
[54] Wu CC, Lee WJ. Synthesis and properties of copolymer epoxy resins prepared from copolymerization of bisphenol A, epichlorohydrin, and liquefied Dendrocalamus latiflorus. J Appl Polym Sci. 2010;116:2065–2073.
[55] Zhang W, Zhang Y, Zhao D. Synthesis of liquefied corn barn-based epoxy resins and their thermodynamic properties. J Appl Polym Sci. 2012;125:2304–2311.
[56] Suzuki M. Activated carbon fiber: Fundamentals and applications. Carbon. 1994;325:77–86.
[57] Ryu Z, Zheng JT, Wang MZ. Porous structure of PAN-based activated carbon fibers. Carbon. 1998;36:427–432.
[58] Kadla JF, Kubo S, Venditti RA, Gilbert RD, Compere AL, Griffith W. Lignin-based carbon fibers for composite fiber applications. Carbon. 2002;40:2913–2920.
[59] Xiaojun M, Guangjie Z. Preliminary study on preparation of carbon fiber from wood–phenol liquefaction products. Chem Ind Forest Prod. 2007;27:29–32.
[60] Li N, Eichhorn SJ. Potential stiffness of carbon fibers produced from highly crystalline cellulose. J Mater Sci. 2006;41:4993–4995.
[61] Sudo K, Shimizu K. New carbon fiber from lignin. J Appl Polym Sci. 1992;42:127–134.
[62] Peng S, Shao H, Hu X. Lyocell fibers as the precursor of carbon fibers. J Appl Polym Sci. 2003;90:1941–1947.
[63] Liu W, Zhao G. Effect of temperature and time on microstructure and surface functional groups of activated carbon fibers prepared from liquefied wood. BioRes. 2012;7:5552–5567.
[64] Li DN, Ma XJ. Preparation and characterization of activated carbon fibers from liquefied wood. Cellulose. 2013;20:1649–1656.
[65] Huang Y, Zhao G. Preparation and characterization of activated carbon fibers from liquefied wood by KOH activation. Holzforschung. 2016;70:195–202.

[66] Huang Y, Ma E, Zhao G. Thermal and structure analysis on reaction mechanisms during the preparation of activated carbon fibers by KOH activation from liquefied wood-based fibers. Ind Crop Prod. 2015;69, 447–455.
[67] Senthilkumaar S, Varadarajab PR, Porkodi K, Subbhuraam CV. Adsorption of methylene blue onto jute fiber carbon: kinetics and equilibrium studies. J Colloid Interface Sci. 2005;284:78–82.
[68] Ma X, Zhao G. Structure and performance of fibers prepared from liquefied wood in phenol. Fiber Polym. 2008, 9, 405–409.
[69] Yoshioka M, Katsura H, Shiraishi N. Fabrication and characterization of carbon fiber derived from woody biomass. J Soc Mater Sci. 2011;60:35–40.
[70] Kunaver M, Jasiukaitytė E, Čuk N, Medved S, Oprešnik SR, Katrašnik T. Biomass waste – a source of raw materials and new energy source. World Renewable Energy Congress, Linköping, Sweden; 2011. p. 239–46.
[71] Kunaver M, Jasiukaityte E, Cuk N, Guthrie JT. Liquefaction of wood; synthesis and characterization of liquefied wood polyester derivatives. J Appl Polym Sci. 2010;115:1265–1271.
[72] Cheumani-Yona AM, Budija F, Hrastnik D, Kutnar A, Pavlic M, Pori P, Tavzes C, Petric M. Preparation of two-component polyurethane coatings from bleached liquefied wood. BioRes. 2015;10:3347–3363.
[73] Budija F, Tavzes Č, Zupančič-Kralj L, Petrič M. Self-crosslinking and film formation ability of liquefied black poplar. Bioresour Technol. 2009;100:3316–3323.
[74] Okabe T, Saito K, Hokkirigawa K. The effect of burning temperature on the structural changes of woodceramics. J Porous Mat. 1996;2:215–221.
[75] Hirose T, Fan TX, Okabe T, Yoshimura M. Effect of carbonization temperature on the basic properties of woodceramics impregnated with liquefied wood. J Mater Sci. 2001;36:4145–4149.
[76] Hirose T, Zhao B, Okabe T, Yoshimura M. Effect of carbonization temperature on the basic properties of woodceramics made from carbonized bamboo fiber and liquefied wood. J Mater Sci. 2002;37:3453–3458.
[77] Zhao BY, Hirose T, Okabe T, Zhang D, Fan TX, Hu KA. Woodceramics prepared from wood powder/phenolated wood composite. J Porous Mat. 2002;9:195–201.

Prasanta Sahoo, Suman Kalyan Das, and Samir Kumar Acharya

2 Tribological aspects of natural fiber composites

Abstract: Composites have become an important engineering material especially in applications requiring low weight combined with decent structural strength. Due to environmental concerns there have been efforts lately to replace the synthetic reinforcing agents with natural products viz. jute, sisal, hemp, bamboo, and so on. Availability of fibrous derivatives from the local vegetation has also driven many researchers to try out a variety of natural fiber reinforced composites. Many of these composites are found to match the properties of the synthetically reinforced composites. Studies related to the tribological characteristics viz. friction and wear of the natural fiber composites have also been addressed. The present chapter attempts to present these tribological results in one place so that the true potential of these composites as tribomaterial can be realized.

2.1 Introduction

Environmental awareness today motivates researchers worldwide in their quest for developing environmental friendly solutions in every aspect of life. The field of tribology is not an exception where naturally derived products are tested for their efficiency to provide the desired friction and wear properties as compared to conventional materials, many of which are not ecofriendly.

For various reasons there is an ongoing process of replacement of metal parts by polymer-based materials. More mileage in cars and airplanes due to lighter weight is one of the desired criteria which can be achieved through the use of polymer-based composites. For example in 2007, Boeing unveiled its 787 Dreamliner, its most fuel efficient aircraft to date, which is constituted of almost 80 % by volume of composite material. In the conception of composite materials, the first and primary aim is to tailor the material properties through the control of fiber–matrix combinations and the selection of processing techniques. An appropriate selection of matrix and the reinforcing phase impart better strength and modulus to a composite comparable to conventional metallic material [1].

The availability of natural fibers and ease of manufacturing have tempted researchers to try out locally available inexpensive fibers and to study their feasibility for reinforcement purposes and to what extent they satisfy the required characteristics of a good reinforced polymer composite for various applications. Due to its low cost and high specific mechanical properties, natural fiber has already displayed its potential to be a good renewable and biodegradable replacement for the most common synthetic reinforcement, i.e., glass. Hence, especially for agricultural states, the green

https://doi.org/10.1515/9783110435788-003

composites not only provide ecofriendly alternatives to various applications but are also a source of revenue by exporting to other countries.

2.2 Significance of tribology in development of materials

Tribology is the study of the science and technology of interacting surfaces in relative motion [2]. It includes the study and application of the principles of friction, lubrication and wear, which are of enormous practical importance, because the functioning of many mechanical, electromechanical, and biological systems depends on the appropriate friction and wear values. In recent decades, the study of tribology has received increasing attention as it has become evident that the wastage of resources resulting from high friction and wear makes up a significant share of the Gross National Product of any country. Hence, the potential savings offered by improved tribological knowledge are immense.

The field of tribology has been progressed to such an extent that laboratory-based experimental methods are available which can evaluate the friction and wear behavior of different materials. This enables researchers to develop newer materials or surface finishing techniques and assess the performance of the same at their convenience and within a very short time. Due to the advancement of this avenue of experimental tribology, commonly used engineering materials as well as advanced multifunctional materials have been tested and people have a fair idea of the tribological characteristics of a variety of materials. However, with the focus shifting towards saving nature, more and more potential environment friendly materials are being developed which are gradually replacing conventional materials. This warrants the need for proper estimation of the tribological performance of these materials so that they can be suitably modified (if needed) to evolve into a high performing material.

2.3 Natural fiber composites

Nature is full of examples wherein the idea of composite materials is used. The coconut leaf is a cantilever using the concept of fiber reinforcement. Wood is a fibrous composite: cellulose fibers in a lignin matrix. The cellulose fibers have high tensile strength but at the same time are very flexible. The lignin matrix on the other hand joins the fibers and provides the stiffness. Bone is yet another example of a natural composite that supports the weight of various parts of the body.

Natural fiber composites, also known as green composites, have received a growth in interest because of their recyclability, biodegradability and abundant availability. The properties such as relatively low density and ability to be tailored to have stacking sequences provide high strength and stiffness in the direction of high loading, which makes these materials attractive compared to conventional metallic systems [3].

Fiber reinforced polymer composites have received considerable attention in various applications because of their suitable properties and superior advantages. Natural fibers are preferred over synthetic fibers in terms of their relatively low weight, low cost, less damage to processing equipment, good mechanical properties such as tensile modulus and flexural modulus, improved surface finish of molded parts. Besides, these materials are biodegradable, have minimal health hazards and are abundantly available. Moreover, locally available vegetation can be used to develop suitable composite materials at a lesser cost. Green composites possessing high strength and stiffness can be produced by adding the tough and light-weight natural fiber into polymer (thermoplastic and thermoset). On the other hand, natural fiber composites are not entirely free from deficits. The natural fibers' structure consists of cellulose, hemicelluloses, lignin, pectin, and waxy substances and permits moisture absorption from the surroundings which causes weak bindings between the fiber and polymer. Furthermore, the couplings between natural fiber and polymer are considered a challenge in effective load transfer, because the chemical structures of both fibers and matrix are different. Hence, these composites require some specific treatments to reduce these deficiencies. These modifications are based on the utilization of reagent functional groups which have the ability to respond to the fiber structures and change their composition. As a result, fiber modifications cause reduction of moisture absorption of the natural fibers which leads to an excellent enhancement incompatibility between the fiber and the polymer matrix.

2.3.1 About natural fibers

In simple terms, fibers that are not synthetic or manmade are natural fibers. These fibers can be sourced from plants or animals. The use of natural fibers as reinforcement for polymer-based composites gained considerable attention in the last decade as is evident from publications statistics as illustrated in Fig. 2.1. Fibers derived from plants such as oil palm, sisal, flax, and jute have been used to produce composite materials that have been very popular and also very well marketed.

2.3.2 Classification of natural fibers

There are three main categories of natural fibers viz. vegetable fibers, animal fibers and mineral fibers. Plant-based natural fibers (vegetable fibers) have high fiber strength and are hence preferred by the automotive industry. The classification of reinforcing natural fibers indicating types of plant and fiber groups is shown in Tab. 2.1. These fibers are renewable, nonabrasive to process equipment and can be incinerated at the end of their life cycle for energy recovery as they possess a good deal of calorific value.

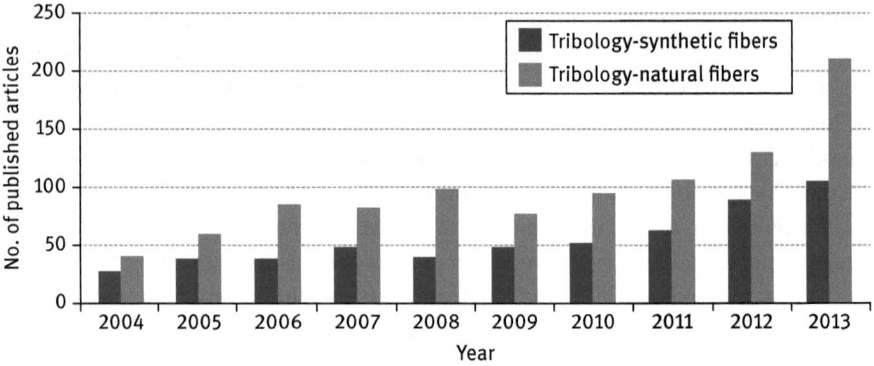

Fig. 2.1: Number of published articles based on tribology of synthetic and natural fibers [4].

Tab. 2.1: Classification of natural fibers used as reinforcements [4].

Wood fiber	**Non-wood fiber**				
	Fruit fiber	**Stalk fiber**	**Bast fiber**	**Seed fiber**	**Leaf fiber**
Hardwood	Coconut	Bamboo	Rattan	Cotton	Sisal
Softwood	Betel nut	Wheat	Hemp	Oil palm seed	Manila
Sawdust		Rice	Jute	Kapok	Banana
		Grass	Ramie	Alfalfa	Palm
		Barley	Banana		Mengkuang
		Corn	Flax		Date palm
			Sugarcane		Pineapple
			Kenaf		Abaca
			Roselle		

They are also much safer to handle and less suspected of affecting the lungs during processing and use.

In the present scenario, natural fibers have excellent potential to reduce not only CO_2 emissions but also save nonrenewable resources by substituting glass fiber reinforcements in plastic composites. Traditionally, glass fibers/wool have been extensively used as building insulation material and reinforcement in auto sector thermoplastics. However, the environmental performance of glass fiber material thermoplastics has several drawbacks due to extensive energy consumption and potential health risks during production and handling. Glass fibers cause severe abrasion to process equipment and their composites may transform into sharp splints during collision causing extra injuries to passengers. Moreover, glass fiber material thermoplastics are nonrecyclable and their incineration generates a clinker-like mass that is hard to dispose of except in landfill [5]. Fig. 2.2 shows some of the popularly used natural fibers.

Fig. 2.2: Some commonly used fibers in green composites for engineering applications (source: Wikimedia Commons, Pixabay, Public domain pictures).

2.3.3 Classification of green composites

An extensive classification for green composites is a bit difficult and also not readily available as there are a number of natural fibers available and newer ones are constantly being added to the list. However, by judgment it seems fairly logical to roughly classify a green composite based on its main constituent viz. the natural fiber and the polymer matrix. A detailed categorization of the natural fibers can be found in the preceding section whereas the main variety of polymers available is given in Tab. 2.2. However, browsing through the literature, it seems fairly logical to classify the natural fiber composites according to the type of reinforcement configuration (illustrated in Fig. 2.3):

- Fiber reinforced: Fibers can be long or short and unidirectional or random
- Mat fiber reinforced: Fibers are woven in the form of mats and arranged in layers
- Flake reinforced: Reinforcement is in the form of flakes.
- Particulate reinforcement : Reinforcement is in the form of small particulates

Apart from the reinforcement type, a simple broad classification of natural fiber composites is given in the following subsections.

Tab. 2.2: Types of polymers and their main applications [4].

Polymer types		**Applications**
Thermosets	Polyester	Composite material/bearing application
	Epoxides	Boat hulls/composite material
	Polyurethanes	Refrigerators
	Phenolics	Gears/bearings/electrical insulators
	Polyimides	Circuit boards/airplane body
Thermoplastics	Acrylics	Name tags/canopies/lenses for car lights
	Polyethylene	Ballpoint tubing/wire & cable insulation
	Polypropylene	Containers/fans/car interior panels
	Polyvinyl chloride	Piping/floor tiles
	Polystyrene	Cups/vending machines/packaging
Elastomers	Natural rubber	Hoses/belts/matting/gloves
	Silicone rubber	Adhesive bonds/O-rings
	Polybutadiene	Automobile tires
	Epichlorohydrin	Explosives/water purification
	Polyether block amide	Damping system components/shoes/insulators

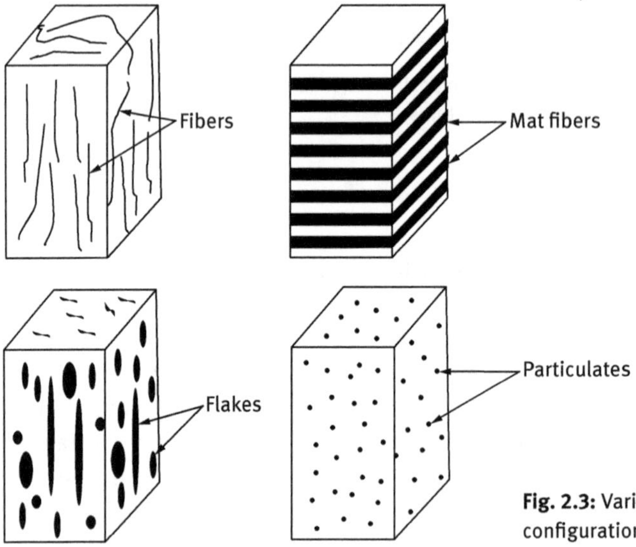

Fig. 2.3: Various types of reinforcement configurations [4].

2.3.3.1 Pure composites

Pure composites contain a single natural type of fiber reinforcement in the matrix. These composites are naturally easier to develop and have fairly good mechanical and tribological properties. However, the composite suffers from the limitations of the properties of the contained fiber. Examples include jute-epoxy composites, kenaf-epoxy composites, etc.

2.3.3.2 Hybrid composites

Although pure natural fiber composites display good mechanical and tribological characteristics, they suffer from some drawbacks due to the inherent limitations of the natural fibers viz. lower mechanical properties (tensile and flexural) as compared to traditional fibers as well as poor resistance to moisture absorption. Hence, use of one natural fiber alone in polymer matrix is inadequate in satisfactorily tackling all the technical needs of a fiber reinforced composite.

A hybrid composite is a combination of two or more different types of fibers in which one type of fiber balances the deficiency of the other. As natural fibers are lower on mechanical strength, synthetic fibers viz. glass are introduced as the second reinforcement so as to best take advantage of the properties of both the fibers. Fig. 2.4 shows a hybrid composite made of jute and glass where the reinforcement is in a mat-type configuration.

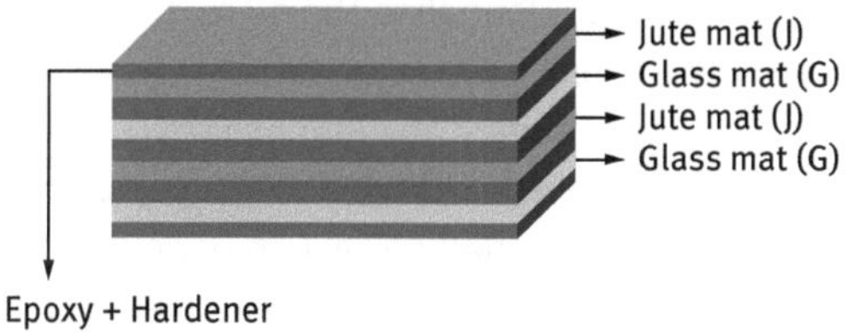

Fig. 2.4: Schematic view of a jute-glass hybrid composite [6].

2.3.3.3 Filler-based: Micro- and nanoparticles

In fiber reinforced polymer composites, fillers may be added to the polymeric matrix for one or more of the following reasons (i) cost reduction, (ii) increase of modulus, (iii) control of viscosity and (iv) production of a smoother surface. Micro- and nanoparticles are presently considered as high-potential filler materials for improving the mechanical and physical properties of polymer composites. Particles like silica (SiO_2), titanium dioxide (TiO_2), calcium carbonate ($CaCO_3$), or polyhedral oligomeric silsesquioxane (POSS), etc., are the commonly used inorganic fillers. However, environmental awareness has attracted researchers to develop new composites with addition of more than one reinforcement from natural resources, such as natural fiber/natural fiber or natural fiber/nanofiller from organic sources as an alternative to synthetic fibers.

2.3.4 Applications of natural fiber composites

Natural fiber-based composites have evolved due to environmental concerns and have shown promising mechanical characteristics. However, due to issues like geographical variation, etc., consistency in fiber properties is a concern as these affect the property of the composite in totality. However, with advancements in processing technologies, these limitations can be minimized to a great extent. The development of hybrid composites is also another step towards popularizing the industrial application of green composites.

Natural fiber reinforced composites are emerging very rapidly as potential substitutes for metal or ceramic-based materials in applications that also include automotive, aerospace, marine, sporting goods and electronic industries [7]. In the US alone, straw is utilized to make composite materials for construction. The automobile industry is at present one sector where the use of naturally-derived composites is very popular due to their high fiber strength. Many automotive components are already produced with natural composites, mainly based on polyester or polypropylene and fibers like flax, hemp, or sisal. Leading German automobile manufacturers, viz. Mercedes, BMW, Audi and Volkswagen have taken the initiative in introducing natural fiber composites for interior and exterior applications. The first commercial example is the inner door panel of the 1999 S-Class Mercedes-Benz, made of 35 % Baypreg F semi-rigid (polyurethane) elastomer from Bayer and 65 % of a blend of flax, hemp and sisal [8]. Mercedes-Benz also used an epoxy matrix with the addition of jute in the door panels in its E-class vehicles as early as 1996. In 2000, Audi launched the A2 midrange car whose door trim panels were made of polyurethane reinforced with a mixed flax/sisal material. Toyota developed an eco-plastic made from sugar cane and use it to line the interiors of the cars [9]. Plastic/wood fiber composites are being used in a large number of applications in decks, docks, window frames and molded panel components [10]. Kenaf/glass epoxy composite materials have also been used to manufacture passenger car bumper beam [11]. Recently, banana fiber reinforced composites are drawing interest due to the innovative application of banana fiber in underfloor protection for passenger cars [12]. Automobile parts such as rearview mirrors, visors in two wheeler, seat covers, indicator covers, side covers, and number plates are fabricated using sisal and roselle fiber hybrid composites.

Green composites have also been applied in the field of biomedicine. The coir/polyester composites have been used to produce mirror casings, paper weights, projector covers, voltage stabilizer covers, mail-boxes, helmets and roofs. In structural infrastructure applications, natural fiber composites have been used to develop load-bearing elements such as beams, roofs, multipurpose panels, water tanks and pedestrian bridges. Jute-based green composites could be suitable for even primary structural applications, such as indoor elements in housing, temporary outdoor applications like low-cost housing for defense and rehabilitation, and transportation. Due to its insulating characteristics, jute may find areas of application in automotive door/ceiling panels and panels separating the engine and passenger compartments.

2.4 Chemical modification of fiber surface-improving fiber-matrix adhesion

Despite the advantages, the performance of untreated natural fiber composites has been much below their capability due to reasons like high moisture absorption of the cellulose fibers, biodegradability of the fibers, limited thermal stability of the composite and most importantly the poor adhesion between the cellulose fibers and polymer matrix. Hence, natural fiber reinforced polymer composites inherently have poor mechanical properties which is also reflected in their tribological performance. Poor adhesion between fiber and substrate renders the composite prone to higher levels of wear and tear when slid against a harder surface.

The interfacial fiber-matrix adhesion can be improved by modifying the fibers, the matrix or both the fibers and the matrix by different physical and chemical methods. Matrix modifications generally involve the addition of chemical coupling agents and compatibilizers to polymer matrix, with the purpose of improving polymer reactivity and wetting the reinforcing fibers. Fiber treatments may be biological, physical or chemical, and are performed to achieve one or more of the following objectives:

- roughening of the fiber surface
- removal of undesirable fiber constituents
- modification of the chemical nature of the fiber surface
- separation of individual fibers from their fiber bundles
- reducing the hydrophilicity of the fibers

Fiber treatments involving the use of chemicals play an important role in improving the reinforcing capabilities of fibers (Fig. 2.5). These treatments can either be classified as fiber pretreatments, coupling agents, compatibilizers or dispersing agents. Pretreatments involve the use of chemicals that remove undesirable and nonstrength-contributing fiber constituents such as lignin, pectin and hemicelluloses. Compatibilizers are chemicals that lower the surface energy of fibers to make them more nonpolar and therefore more compatible with polymer matrices. Dispersing agents are used to improve the dispersion of fibers in the matrix. Coupling agents are mainly responsible for improving the adhesion between reinforcing fibers and the matrix material, but can also reduce the water uptake of the fibers and assist with fiber dispersion as well. Due to this overlap in functions and to simplify matters, all bonding agents and surfactants have been grouped together as chemical treatments.

At present, over forty coupling agents have been used in the production and research of natural fiber composites [13]. The most popular treatments include the use

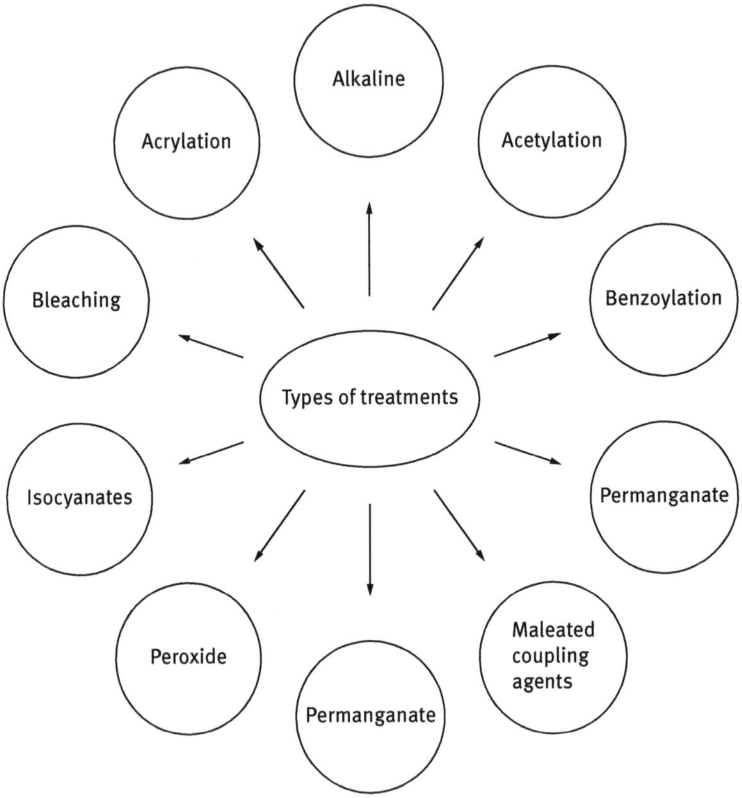

Fig. 2.5: Different types of treatments on natural fibers [4].

of alkalis, anhydrides and anhydride-modified copolymers, benzoyl chloride, permanganate silanes and isocyanates.

2.4.1 Alkaline treatment

Alkaline treatment is one of the most used chemical treatments of natural fibers that are used to reinforce thermoplastics and thermosets. The important modification done by alkaline treatment is the disruption of hydrogen bonding in the network structure, thereby increasing surface roughness. This treatment removes a certain amount of lignin, wax and oils covering the external surface of the fiber cell wall, depolymerizes cellulose and exposes the short length crystallites [14]. Addition of aqueous sodium hydroxide (NaOH) to natural fibers promotes the ionization of the hydroxyl group to the alkoxide [15].

$$\text{Fiber–OH} + \text{NaOH} \longrightarrow \text{Fiber–O–Na} + H_2O \tag{2.1}$$

Thus, alkaline processing directly influences the cellulosic fibril, the degree of polymerization and the extraction of lignin and hemicellulosic compounds [16]. It is reported that alkaline treatment has two effects on the fiber:

1. It increases surface roughness resulting in better mechanical interlocking, and
2. It increases the amount of cellulose exposed on the fiber surface, thus increasing the number of possible reaction sites [17].

Consequently, alkaline treatment has a lasting effect on the mechanical behavior of flax fiber, especially on fiber strength and stiffness.

2.4.2 Benzoyl chloride treatment

Benzoylation is an important transformation in organic synthesis. Benzoyl chloride is most often used in fiber treatment. Benzoyl chloride includes benzoyl ($C_6H_5C{=}O$) which contributes to the decreased hydrophilic nature of the treated fiber and improved interaction with the hydrophobic polymer matrix. The reaction between the cellulosic hydroxyl group of the fiber and benzoyl chloride is given as follows:

$$\text{Fiber—OH} + \text{NaOH} \longrightarrow \text{Fiber—O}^-\text{Na}^+ + \text{H}_2\text{O} \tag{2.2}$$

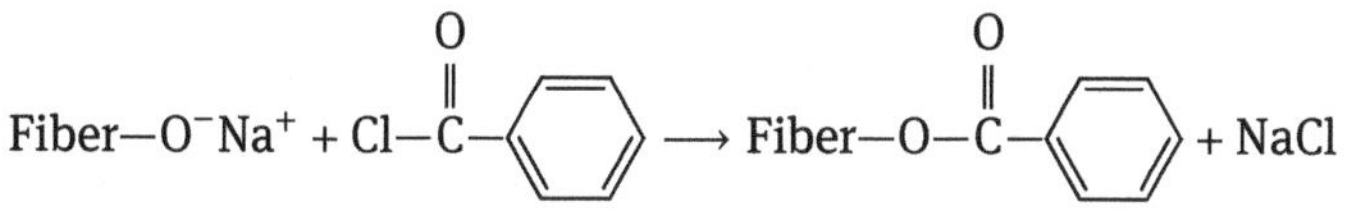
(2.3)

Benzoylation of fibers improves fiber-matrix adhesion, thereby considerably increasing the strength of the composite, decreasing its water absorption and improving its thermal stability [18].

2.4.3 Permanganate treatment

Permanganate treatment on natural fibers is generally conducted with potassium permanganate ($KMnO_4$) in acetone solution. This treatment forms highly reactive permanganate (Mn^{3+}) ions that react with the cellulose hydroxyl groups and form cellulose manganate for initiating graft copolymerization. This treatment enhances chemical interlocking at the interface and provides better adhesion with the matrix [18]. Formation of cellulose manganate is responsible for higher thermal stability of the fiber. It also reacts with the lignin (hydrophilic –OH groups) constituents and separates from the fiber cell wall. It reduces the hydrophilic nature of the fiber. Higher concentrations of $KMnO_4$ (more than 1 %) cause excess delignification (removal of cementing materials) within the cellulosic structure and degrade fiber properties [19].

Paul et al. [19] reported that, during an oxidation reaction, $KMnO_4$ etches the fiber surface and makes it physically rougher to improve mechanical interlocking with the matrix. Flexural strength and modulus properties increased by 5 % and 10 % for the treated banana fiber polypropylene composites. Li et al. [20] applied 0.2 % potassium permanganate ($KMnO_4$) solution (in 2 % acetone) on alkali (2 % NaOH for 1 h) pre-treated flax fiber and reported treated fiber-LLDPE and HDPE composites had higher tensile strength properties compared to the untreated fiber composites. The reaction between fiber –OH group and potassium permanganate is given in equation (2.4).

$$\text{Fiber—OH} + KMnO_4 \longrightarrow \text{Fiber—O—H—O—}\overset{\overset{\displaystyle O}{\|}}{\underset{\underset{\displaystyle O}{\|}}{Mn}}\text{—OK}^+ \tag{2.4}$$

A number of investigations have been made on various chemical treatments of natural fibers such as kenaf, hemp, flax and jute to enhance the mechanical properties of composite materials.

2.5 Mechanical characteristics of natural fiber composites

Natural fibers are known for their high strength and hence are expected to impart the same to the composites. However, there are many suggested treatments for natural fibers that can be implemented in order to enhance their mechanical properties resulting in high strength and structure. According to some researchers [21], once the base structures are made strong, the polymers can be easily strengthened and improved. There are various factors affecting the performance of composites including [5]:
1. orientation of fibers
2. strength of fibers
3. physical properties of fibers
4. interfacial adhesion property of fibers

The mechanical efficiency of green composites is dependent upon the interface provided by fiber-matrix along with the stress transfer function in which stress is transferred to fiber from matrix. Characteristic components of natural fibers such as orientation, moisture absorption, impurities, physical properties, and volume fraction are such features that play a constitutive role in the determination of mechanical properties of the composites. Mechanical properties of PLA (poly lactic acid), epoxy, polypropylene and polyester matrices can be affected by different types of natural fibers. Improvement of about 75 % in tensile strength is observed when jute fibers are added to PLA. Similarly, improvement is also noticed when hemp, kenaf and cotton fibers are incorporated in polypropylene (PP) matrix [22]. However, introduction of flax fibers showed a negative impact on the tensile strength of PLA which reduced by about 16 % compared to neat PLA. This suggests that the compatibility between the matrix and the fiber is very important in order to achieve good performance from the composite. The maximum improvement in the tensile strength is observed when jute fibers are added to polyester. The improvement is of the order of 121 % compared to pure polyester [22].

It is also known that stiffness and stress transfer in composites increase with an increased or excessive addition of fiber which provides a better loss modulus and also a better storage modulus. On increasing the fiber content, the strength of the composite together with the modulus and fracture strength increases and the best combination is found with 30 volume percentage of fiber. The chemical treatment of the fiber surface significantly improves the fiber-matrix adhesion, which in turn improves the mechanical properties of composites [23]. Rahman et al. [24] evaluated that with increasing the fiber loading, there is a gradual increase in the strength, Young's modulus, flexural modulus, and hardness of jute fiber composite, and 30 % jute fiber loading composite gave the optimum value for mechanical properties. John and Naidu [25] studied the tensile strength of unsaturated polyester-based sisal-glass hybrid composites with fiber loading and found a significant improvement in tensile strength due to NaOH and trimethoxysilane treatment. Mishra et al. [26] developed and compared two different types of composites: glass-sisal and glass-pineapple with polyester as resins and also studied their mechanical properties such as tensile and flexural strength and impact strength.

In this regard, various attempt has been made by different researchers to combine varieties of natural fiber with synthetic fiber. Jose Almeida et al. [27] evaluated the enhancement in the properties of curaua polyester composites by incorporating chopped glass fibers with chopped curaua fiber. Hybrid composite using jute and glass fibers were developed by stacking alternately layers (mats) of each fiber as illustrated in Fig. 2.4 [6]. Each layer is woven from the respective fibers and the composites are prepared by hand lay-up technique. The variation of tensile strength for various laminate stacking sequences of jute-glass fiber epoxy composites is shown in Fig. 2.6. The tensile strength of unreinforced epoxy resin is found to be around 21.03 MPa. When only laminates of jute and glass fibers are reinforced into epoxy, tensile strength is found to be 142 % and 442 % greater than the neat epoxy resin. An increase in the tensile

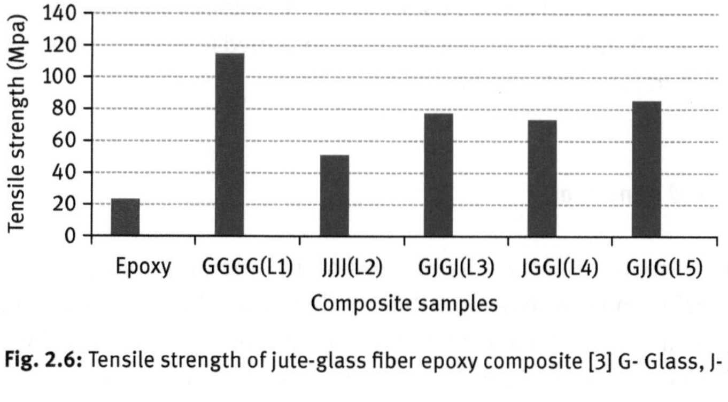

Fig. 2.6: Tensile strength of jute-glass fiber epoxy composite [3] G- Glass, J- Jute.

strength of 51 %, 41 %, and 66 % is observed for 50 : 50 jute-glass fiber reinforced hybrid laminate (L3, L4, and L5) composites when compared to that of only jute laminate (L2) composite (Fig. 2.6). The jute fiber reinforced epoxy composites give about 55 % strength of the glass fiber reinforced epoxy composites. It was also found that there is a sharp increase in the tensile strength of the composites with the incorporation of glass fibers

The bonding strength between fiber and polymer matrix in the composite is considered a major factor in order to get superior fiber composite properties. The presence of pendant hydroxyl and polar groups in the fiber leads to extremely high moisture absorption by the fiber, resulting in weak interfacial bonding between the fiber and the hydrophobic matrix polymers. To develop composites with good mechanical properties, chemical modification of the fiber can be carried out to reduce the hydrophilic behavior of fibers and the absorption of moisture [10].The different surface treatments of advanced composites applications have been reviewed by several researchers [28]. The effects of different chemical treatments on cellulosic fibers that are employed as reinforcements for thermoplastics and thermoset are also examined.

Pavithran et al. [29] determined the work of fracture by impact testing on sisal-glass hybrid composites with two arrangements, one with sisal shell and glass core and the other with glass shell and sisal core. They showed that the sisal shell laminate had the higher work of fracture compared to glass shell hybrid laminates of equivalent volume fraction of sisal and glass fibers. Mishra et al. [26] studied the effect of glass fiber addition on tensile and flexural strength and Izod impact strength of pineapple leaf fiber (PALF) and sisal fiber reinforced polyester composites. K. John et al. [25] studied unsaturated polyester-based sisal glass composites with 5 % and 8 % volume fraction and found a considerable enhancement in impact, compression flexural and tensile properties. Pavithran et al. [30] also prepared coir-glass hybrid polyester composites using hydroforming method and also did a comparison between the moisture-absorption behavior of coir-polyester composites and hybrid composites. The results revealed that the properties of coir fiber composite were enhanced due to the addition of glass fiber.

Kalaprasad et al. [31] studied the mechanical properties of sisal-glass fiber reinforced low-density polyethylene matrix composites by taking into account the effects of fiber orientation. The result showed that longitudinally oriented fibers exhibit better mechanical properties than randomly oriented fibers.

2.5.1 Effect of micro- and nanofiller

A high interfacial area of matrix filler results because of uniform and homogeneous dispersion of nanoparticles and is also responsible for changing relaxation behavior, as well as in ensuing the mechanical, molecular mobility, and thermal properties [32]. Generally, nano-sized filler is present in the minor zone whereas only few of the microparticles participate in the plastic zone deformation. This provides a way for the nanofillers to improve the fracture and mechanical properties of a matrix having a brittle property.

Hybridization involving the combination of nanofiller and natural fiber in the matrix results in reduction of water absorption properties and an increase in mechanical properties [33]. Organic nanohybrid composites have been prepared by various researchers by introducing organically derived fillers viz. coir nanofiller, carbon black, cellulosic nanofiller, etc. into matrices made by combining natural fibers/natural fibers and natural fibers/synthetic fibers with epoxy, polyester, phenolic, poly vinyl ester, poly urethane resins, etc. [34, 35]. Natural fiber/nanofiller-based hybrid composites can be utilized in the building and construction sector, transportation (automobiles, railway coaches, aerospace), packaging, consumer products, etc., and also could possibly produce acoustic insulator and extremely thermally stable materials.

Raghavendra [6] has further studied the effect of the introduction of nanofillers into the epoxy matrix of pure as well as hybrid composites (Fig. 2.7). An increase of about 5 % in the tensile strength of pure and hybrid jute composites is observed by addition of fly ash as microfiller [6]. With an increase in weight percentage of micro fly ash (MFA) the tensile strength of the composite increases (shown in Fig. 2.8). Beyond 10 weight percentage however, the strength shows a decreasing trend [6]. This may be due to improper bonding of the fly ash particles with the matrix material.

With the introduction of fly ash as nanofiller (NFA), again the tensile strength of both pure and hybrid jute composites increase up to 4 wt% and decrease thereafter (Fig. 2.9). The interesting point here is that at 4 wt% of nanofiller reinforcement for GJJG composite, strength almost catches up with the strength of the pure glass fiber composite.

Similarly, studies have also been undertaken with alumina as the filler material. Tensile properties similar to fly ash jute composites are observed in this case. The tensile strength increases with the increase in weight percentage of alumina, but up to 10 %. Beyond 10 %, the tensile strength decreases which may again be due to the higher volume fraction of filler leading to de-bonding of fibers with the matrix. This also leads to nonadherence of filler with the matrix which in turn decreases the load carrying capacity of the composite. All this shows up in the form of lower strength of the composites with higher particle loading [36]. The results of nano alumina filler show almost the same trend for different composites for the strength values (Fig. 2.10).

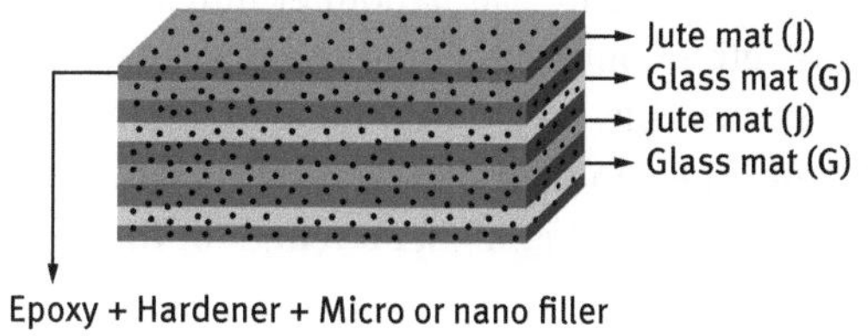

Fig. 2.7: Schematic view of the hybrid nanocomposite.

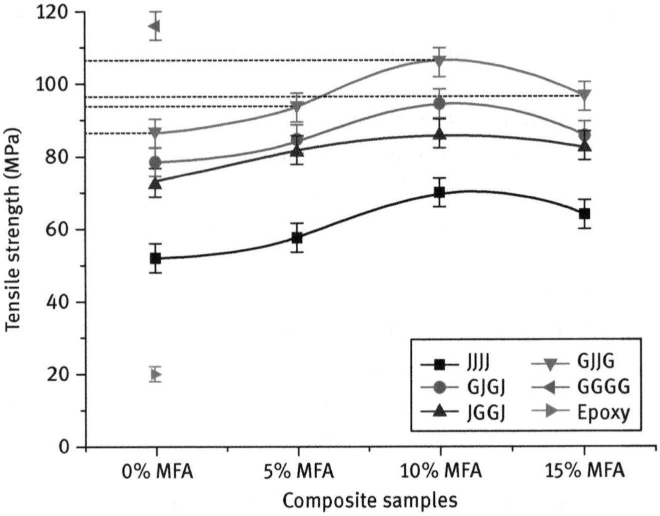

Fig. 2.8: Effect of micro fly ash on tensile strength of hybrid composites (MFA- Micro Fly Ash).

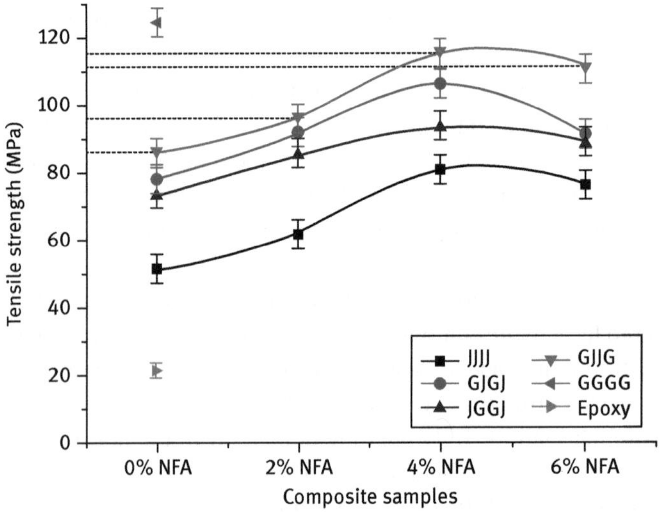

Fig. 2.9: Effect of nano fly ash on tensile strength of hybrid composites (NFA- Nano Fly Ash).

The benefit of alumina nanofiller in comparison to fly ash nanofiller indicates that both tensile strength values cross the strength of pure glass fiber composite. The increase in tensile strength is found to be 5 % for GJJG with 4 wt% nano alumina composite compared to glass only (GGGG) composite. Since alumina is a ceramic filler for which fracture toughness is higher [37], results showing higher strength of the composite with 4 wt% of nanofiller are achieved.

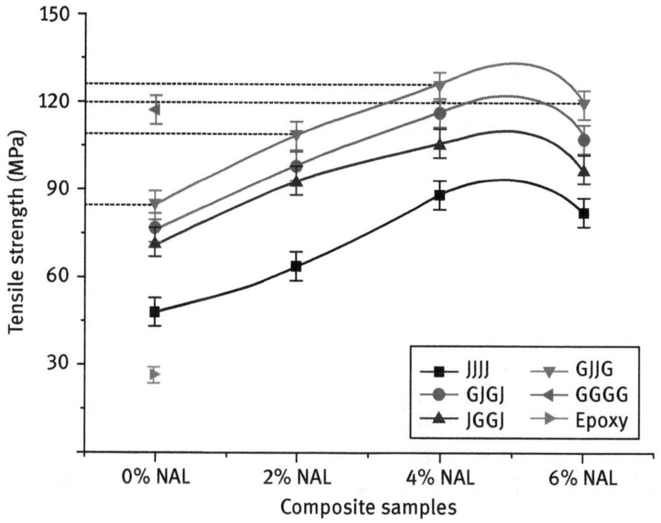

Fig. 2.10: Effect of nano alumina on tensile strength of hybrid composite.

Mechanical and tribological performance of the date palm fiber/epoxy composites are enhanced by addition of graphite filler, but high content of the graphite deteriorates the mechanical properties [32].

2.5.2 Morphology of tensile tested composites

Scanning Electron Microscope (SEM) micrographs of the tensile fracture surfaces of the pure jute, glass, and hybrid with micro- and nanofiller composites are shown in Fig. 2.11 (a)–(e) respectively. From the SEM image (Fig. 2.11 (a)) of raw jute fiber epoxy composites, it is clearly observed that due to tensile load the fiber gets a little stretched and has under gone breakage. This feature suggests weak interfacial bonding between the fiber and matrix. When the glass fiber composite shown in Fig. 2.11 (b) is considered, it is clearly observed that the fiber gets stretched and finally broke after a long elongation. This validates the higher value of tensile strength of the glass fiber composites.

When considering hybrid composites, the tensile strength of the composites is less than compared to the pure glass fiber, which is due to the early breakage of jute fibers. The morphology of the hybrid composite shown in Fig. 2.11 (c) supports this result. From Fig. 2.11 (d), it is clearly observed that the crack propagation in epoxy composite is reduced with fly ash filler addition. From Fig. 2.11 (e), fiber twisting is observed due to tensile load before it undergoes breakage. When compared to Fig. 2.11 (d), this effect is due to alumina filler addition in composites.

Fig. 2.11: SEM image of (a) JJJJ fiber composite, (b) GGGG fiber composite, (c) GJJG hybrid composite, (d) GJJG hybrid composite with nano fly ash filler and (e) GJJG hybrid composite with nano alumina filler [3].

2.6 Tribological behavior of natural fiber composites

Tribology is vital to modern machinery involving sliding and rolling surfaces. From a tribological point of view material should be developed in such a way so that its surface is both smooth and hard. This way both friction and wear are reduced. Among the two degrading phenomena, friction is a critical factor dictating the efficiency of mechanical assemblies that involve sliding surface contact. It is also responsible for wear, which is often the limiting mechanism of device service life. Hence, the tribological loadings should be taken into account for an improved part design.

In the case of composites, the main issue is the heterogeneity of the material which reflects on the surface as well. Hence, surface finish operations may often be required to get the desired smoothness of composite materials. Moreover, improper bonding between the matrix and the reinforcing fibers frequently leads to a high amount of wear in sliding contact.

2.6.1 Measuring friction and wear

For evaluation of the friction and wear characteristics of materials, tribotesting apparatus called tribometers or tribotesters are available which enable one to apply the predetermined load and lubrication, if any, and obtain the friction and wear behavior from the sample. In most tribometers used in the laboratory, test specimens of simple geometry are utilized in order to start the tests with test conditions as well defined as possible. Usually one of the moving specimens consists of a rotating cylinder or disc against which another solid specimen of different curvature is pressed, thus leading to nominal point, line or flat contact. The operating variables of a tribometer are broadly given by the following quantities: type of motion, load applied, velocity, ambient temperature and test duration [38].

Type of motion

The type of motion is connected with the geometry of the test system and is given by one of the four basic motion types, viz. sliding, rolling, spin and impact or by a superposition of these basic types.

Load

The load can be applied by various means to the test system, e.g., as a mass (dead load), by a spring, by hydraulic means or by electromagnetic means. For measuring the load, force transducers based on strain gauges, inductive elements or pressure-sensitive detectors attached to the surface of the specimen in the contact zone are available. Load measurement by means of an electromechanical force transducer has the great advantage that the normal load signal together with the signal of the friction force can be fed into an electronic divider and the actual value of the friction coefficient (μ = friction force/normal load) can be calculated directly during the test and on a real-time basis.

Velocity

For controlling velocity, electronic or optoelectronic transducers are commercially available. Velocity is an important parameter in tribotesting as much as it determines the lubrication regime within the Stribeck curve. It is also the independent variable in the product of friction force and sliding velocity which gives the friction-induced power loss of a tribotesting system.

Temperature

The ambient temperature of the test system is another important operating variable since it determines the thermal state of the system – at least at the initial stage of the test. For controlling temperature, the following two techniques are utilized most often: thermocouple techniques and infrared-pyrometer techniques.

Test duration

The set of operating variables (load, velocity and temperature) acts on the tribotesting system as a function of time or duration. The test duration also should be selected to be consistent with the nature of the investigation. Attempts are frequently made to shorten the test duration in so-called "accelerated tests" by increasing the severity of the operating conditions. This procedure, however, is only permissible if the nature of the tribological processes is not changed.

2.6.2 Friction characteristics of natural fiber composites

In any industry, a huge proportion of the useful energy is consumed by friction. Hence, tribologists around the world try to minimize friction by developing advanced materials or by adopting a suitable surface finish method. In the case of natural fiber reinforced composites, most of the work related to tribology is devoted towards studying their wear behavior and mechanism. However, in a few cases the friction behavior of the composites is deliberated in some detail.

It has been observed that the coefficient of friction of epoxy composite decreases with addition of Lantana camara fiber, which confirms that the addition of this fiber is beneficial in reducing the wear of neat epoxy [23]. Bajpai et al. [39] have developed a laminated composite using three plant fibers (nettle, Grew optiva and sisal) into poly lactic acid (PLA) polymer for tribo-based applications. Poly lactic acid, a fully bio-based polymer, is derived from the fermentation of corn, potato, sugar beet, and other agriculture-based materials. PLA has good mechanical properties, excellent barrier capability and can be easily processed for various applications. It has been observed that the minimum friction coefficient of every composite is slightly higher than that of the neat PLA. But, the maximum friction coefficient of composites is significantly lower than that of the neat PLA. At load and speed combination of 30 N and 3 m/s respectively, PLA-Nettle composite shows much higher friction force than that of the PLA-Grew optiva and PLA-Sisal composites. However, at lower load and speed values, all three composites display similar friction forces. Also, not much difference in friction coefficient is observed between neat PLA and its developed composites. The nature of the friction coefficient (increase, decrease or constant) depends upon the formation of thin polymer film and nature of fibers at the interface [40]. Initially, at low load, an increased value of the friction coefficient is due to mechanical interlocking of asperities at the interface. As the load increases, polymer debris deform plastically and a thin layer of polymer adheres to the composite interface covering the fiber cross section. Thin film acts as a protective layer [41] which reduces the friction coefficient.

2.6.3 Wear behavior of natural fiber composites

Wear is an inevitable companion of friction and is responsible for a great loss of economy. It should be minimized at all cost so that device life is not endangered and replacement costs are minimized. Wear of materials encountered in industrial situations can be grouped into different categories as shown in Tab. 2.3. However, there are also mixed mode situations where one type of wear changes to another or where two or more mechanism play together.

Tab. 2.3: Types of wear in industry [42].

Type of wear in industry	Approximate percentage involved
Abrasive	50
Adhesive	15
Erosion	8
Fretting	8
Chemical	5

In the past few decades, numerous works have been carried out on the wear performance of polymer and polymer-based composites. However, in recent years some works have also been attempted on composites using natural fiber as reinforcement. An overall conclusion from all of these works is that wear resistance of polymers improves when natural fibers are introduced. Incorporation of Lantana camara fiber improves the abrasive wear behavior of epoxy-based composites significantly [43]. The effect of fiber content was investigated and it was found that 40 weight percentage yields the best wear performance in the composite. Increased wear at higher fiber concentration might be due to the agglomeration of fiber and lower interfacial adhesion which may further lead to the drawing out of the fibers from the matrix resin during the test. The abrasive wear rate of the Lantana camara fiber-epoxy composite is influenced by several parameters, e.g., sliding velocity, sliding distance and normal load. The wear rate of the composite was found to be more sensitive to normal load in comparison to sliding velocity. While investigating the effect of applied load on the wear resistance of the composite, Deo and Acharya [44] further found that the weight loss increases with the normal load (Fig. 2.12). Weight loss is relatively low at lower load (5 N) because of lower penetration and because a lower number of abrasive particles are in action with the rubbing surface. Abrasion wear is greatly increased at higher load because most of the abrasive particles penetrate into the surface and create more grooves, resulting in more material removal by plowing. At the same load, the composite displayed increasing wear with higher sliding speeds.

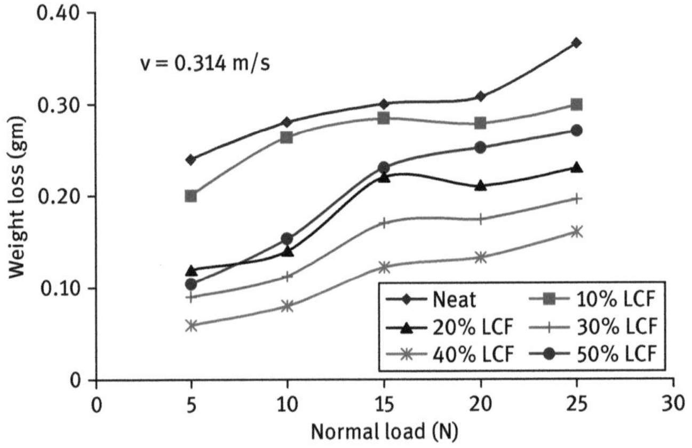

Fig. 2.12: Abrasive weight loss of reinforced and nonreinforced composites vs normal load at constant sliding velocity (03.14 m/s) [44].

At constant applied load (20 N), the specific wear rate decreases with increasing sliding distance for all the samples. Higher wear rate observed initially is due to the fresh abrasive paper. With consecutive runs, the wear rate decreases gradually because the abrasive grits become smooth and less effective. The wear debris fills the space between the abrasives, which reduces the depth of penetration in the sample.

Bajpai et al. [39] observed that wear performance of PLA is significantly improved due to the addition of natural fibers (nettle, Grew optiva and sisal) because the specific wear rate of composites was considerably lower compared to the specific wear rate of neat PLA, especially at higher loads. Neat polymer has poor wear resistance due to the softening process and subsequent film generation and removal process during sliding [45]. However, for various PLA composites developed, no such distortion was observed during the whole experimentation at similar conditions. Specific wear rate of composites is quite low compared to neat PLA. This confirms that natural fiber reinforcement improves the wear behavior of neat PLA. The highest specific wear rate among the three composites developed is associated with PLA/Sisal composite at 1 m/s sliding speed, which is very low compared to the specific wear rate of neat PLA. In the presence of fibers in the matrix, both fiber layers and matrix are rubbed during sliding and thus fibers support the matrix and protect the matrix layer from being removed. This depends upon the interfacial adhesion characteristics at the fiber-matrix interface. It has been reported earlier that the improved wear performance of betel nut reinforced polyester composites was highly controlled by the presence of fiber mat and its orientation [45]. In the presence of fibers in the matrix, both fibers and matrix participate during sliding operation. A good matrix system should transfer and evenly distribute the applied load onto the fibers, which in turn will increase the wear resistance of the composite. This depends upon good interfacial adhesion characteristics at the fiber-matrix interface. The specific wear rate (SWR) increases as applied load increases at different sliding speeds (1–3 m/s) for neat PLA and each type of composite. An increase in specific wear rate with an increase in load implies that as applied load increases, weight loss of specimen increases with a higher rate.

2.6.3.1 Erosion by hard particles

The subject of erosion wear of polymer composite has received substantial attention in the past decades. Interest in this area is commensurate with the increasing utilization of polymer-based composites in aerospace, transportation and processes industries, where they can be subjected to multiple solid or liquid particle impact. Examples of such applications are pipelines carrying sand slurries in petroleum refining, helicopter rotor blades, pump impeller blades, high speed vehicles and aircraft operating in desert environments, radomes, surfing boats where the components encounter impact of many abrasives like dust, sand, splinters of materials, slurry of solid particles and consequently the material undergoes erosive wear [46]. The most important factors influencing the erosion rate of the composite materials can be summarized under four categories [47]:

1. The properties of the target materials (matrix material properties and morphology, reinforcement type, amount and orientation, interface properties between the matrices and reinforcements, etc.).
2. Environment and testing conditions (temperature, chemical interaction of erodent with the target).
3. Operating parameters (angle of impingement, impinging velocity, particle flux-mass per unit time, etc.).
4. The properties of the erodent (size, shape, type, hardness, etc.).

Thus, it seems that the erosion resistance of the material can be evaluated after investigating the combination of above parameters. In general, erosive behavior of materials

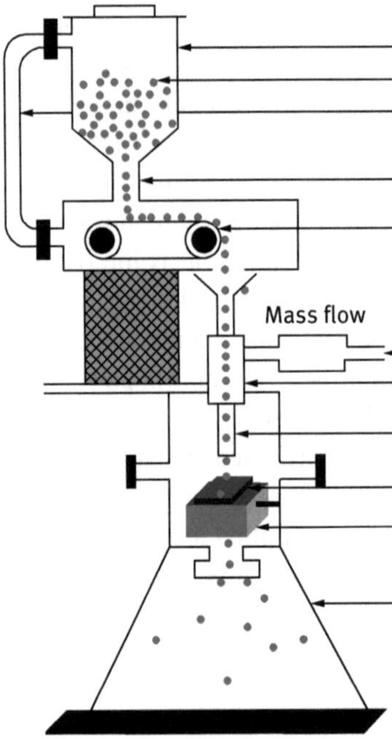

Fig. 2.13: Schematic diagram of erosion test rig [6].

can be grouped into ductile and brittle when the erosion rate is evaluated as a function of impact angle. Ductile behavior is characterized by maximum erosion at low impact angle in the range of 15–30°. On the other hand, if maximum erosion occurs at 90°, then the behavior can be termed brittle. Reinforced composites have also sometimes been found to exhibit an intermediate behavior known as semi-ductile with maximum erosion occurring at an angle in the range of 45–60° [48].

The schematic diagram of the erosion test setup [6] designed as per ASTM-G76 standard is shown in Fig. 2.13. The rig consists of an air compressor, a particle feeder, and an air particle mixing and accelerating chamber. The compressed dry air is mixed with the erodent particles, which are fed at a constant rate from a conveyor belt-type feeder into the mixing chamber and then accelerated by passing the mixture through a tungsten carbide converging nozzle. These accelerated particles impact the specimen, and the specimen could be held at various angles with respect to the impacting particles using an adjustable sample holder. The test apparatus is also fitted with a rotating double disc to measure the velocity of the erodent particle. The loss of weight suffered by the sample represents erosive wear.

Mishra and Acharya [49], Deo and Acharya [50], and Gupta et al. [51] have reported the tribological potential of sugarcane, Lantana camara and bamboo fiber reinforcement in thermoset polymers for enhancing erosive wear resistance. They have the opinion that fiber volume fraction has a significant influence on the erosion rate. Mohanty et al. [52] studied the erosive wear behaviors of date palm leaf fiber reinforced polyvinyl alcohol composites and wood apple shell fiber reinforced epoxy composites. It is reported that [34] if natural fiber is hybridized with a synthetic fiber in the same matrix, the properties of the natural fiber could be improved by taking advantage of both the fibers.

Raghavendra [6] investigated the solid particle erosion behavior of epoxy-based jute and glass hybrid composites. He observed that jute fiber/glass epoxy hybrid composite for all sequences shows peak erosion rate at 60° impact angle, except jute and neat epoxy which show peak erosion at 90° impact angle. Minimum erosion rate for all composites along with pure epoxy is found at an angle of 30° under any impact velocity. Generally, it has been recognized that peak erosion exists at low impact angles (15–30°) for ductile materials and at a high impact angle (90°) for brittle materials [53]. However, the maximum erosion occurring in the angular range 45–60° indicates the semi-ductile behavior of the composite material. A similar type of behavior is also reported by Biswas et al. [54] while studying the erosive behavior of red mud filled bamboo-epoxy composite. It is further noticed that irrespective of impact velocity and impact angle, the erosion rate is highest for neat epoxy and is lowest for jute fiber reinforced epoxy composite. Raghavendra [6] also found that layering sequence and velocity of impact have a significant influence on the erosion rate of the hybrid composite. The morphologies of the eroded surfaces (Fig. 2.14) or the hybrid composite suggest that overall erosion damage of the composite is mainly due to breaking of fibers. Chipping out of fibers is prevented due to good bonding between the fiber and the matrix. Finally, it is concluded that erosive strength of pure jute composite is higher in comparison to hybrid composite.

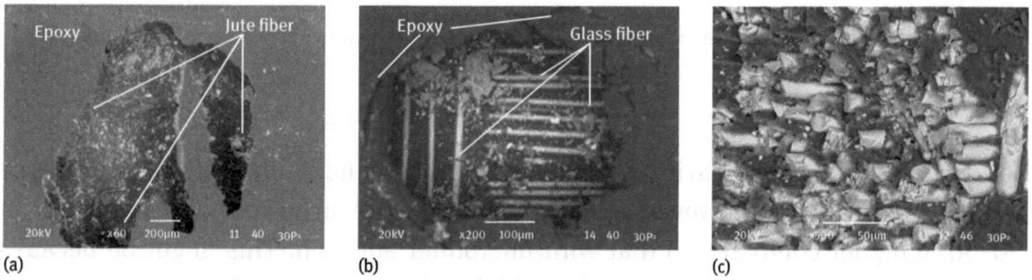

Fig. 2.14: (a) Jute fiber composites at 60° impingement angle, (b) glass fiber composite at 60° impingement angle [55], and (c) hybrid fiber composite at 60° impingement angle [6].

2.6.3.1.1 Effect of fillers on erosive wear

When pure jute, hybrid jute and glass composites are included with micro- and nanofillers, it is found that there is an increase in the erosion resistance of the composites [6]. It is clearly observed from these plots that addition of 5 wt% of filler did not change the behavior of the composite from semi-brittle nature because maximum erosion occurs at 60°. With further addition of filler at 10 and 15 wt%, the behavior of the composite changes from semi-brittle to brittle as maximum erosion occurs at 90° [56]. Jute fiber as the outer layer in a hybrid composite exhibits pure brittle nature, whereas glass fiber as outer layer exhibits semi-brittle nature. Interestingly, the pure jute layered composite with fly ash filler addition shows the best wear resistance when compared to the hybrid and glass fiber composite with and without filler addition [55]. This peculiar behavior of the jute fiber is due to the absorption and bonding behavior of the jute outer surface structure, which indeed leads to great wetting and mixing ability of the jute with the epoxy matrix when the epoxy reacted with the hardener. The presence of hydroxyl ions leads to this type of characteristics in jute and epoxy. Moreover, Fig. 2.15 (a) and (b) proves that jute has a more bonding nature compared to glass fiber. These figures clearly show the traces of epoxy in the jute fiber whereas no traces of epoxy are found with the glass fiber after the test. This might be the cause of the higher impact resistance of jute fiber composites with micro- and nanofillers.

(a) (b)

Fig. 2.15: (a) Microscopic image of glass fiber in epoxy and (b) microscopic image of jute fiber in epoxy [55].

It was found that the behavior of the composites with nanofillers is similar to those with microfillers [6]. However, the erosion resistance achieved with nanofiller is 20–30 % higher compared to that with microfiller addition. This might be because the fly ash and alumina nanoparticles which have a higher surface area and lower particle size fill the gaps that are present in the jute fiber and create a mechanical doping there by making the jute fiber behave like continuous fibers. This creates a good interfacial adhesion between fibers and matrix and the fiber sustains micro-damage caused by impacting particles. Dasari et al. [57] reported a similar type of observation in their work on wear/scratch damage in polymer nanocomposites.

The alumina filler increases the wear resistance of hybrid jute composites to reduce the erosion rate. The formation of cracks due to sand impact supports this statement. Higher magnification micrographs reveal that solid sand particles impact on the jute fibers and cause the fibers to break owing to the formation of cracks perpendicular to their length, but fiber chip out is resisted by the nanofillers. Even when nanoparticles of alumina are introduced into the composite, similar fiber breakage and detaching of the fiber is observed for glass fiber. It is also clearly observed from a microstructure study that due to filler addition, material removal is less and due to the jute hybridization, the bonding between the fiber and the matrix is also increased. When the damage enters the glass fiber layers, it breaks the fibers, but chipping out of fibers (Fig. 2.16) from the matrix is not visible.

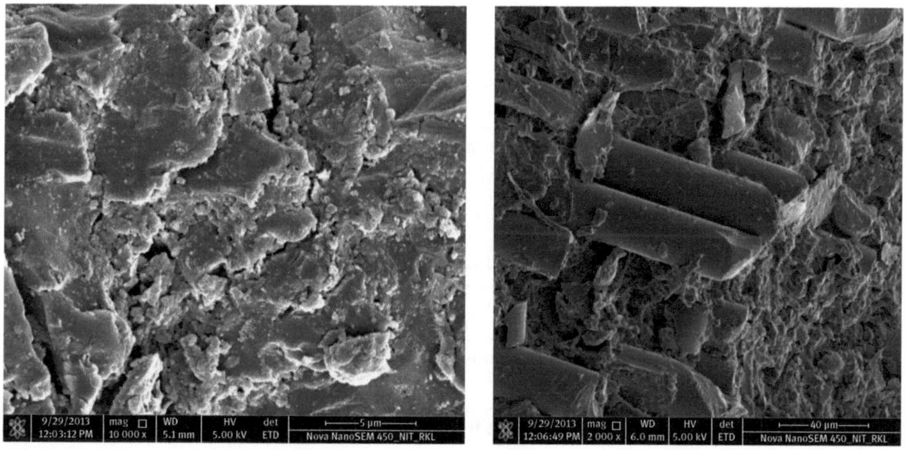

(a) (b)

Fig. 2.16: (a) Jute fiber composite with nano alumina filler at 90° impact angle and (b) glass fiber hybrid composite with nano alumina filler at 90° impact angle [55].

2.6.4 Post-test morphology and wear mechanism

There are several wear mechanisms and each one leaves its distinct signature on the tested specimen. By carefully examining the wear marks, one can estimate the wear mechanism fairly accurately. Even mixed modes of wear can also be identified. Appreciating the various modes of wear: adhesive, delamination, fretting, abrasive, erosive, corrosive, oxidational (mild and severe), melt, the wear mechanism maps enable the researcher to explain the physical interpretations of the wear behavior of a particular material. A summary of the appearance and symptoms of different wear mechanisms is indicated in Tab. 2.4 which is also a systematic approach to diagnose the wear mechanisms.

For PLA-based composites, significant debonding is observed in SEM micrographs of composite specimens after sliding wear testing [39]. The images of the wear zone also showed plowing in the resinous region confirming the presence of abrasive type of wear mechanism. Results show that the wear performance of the developed composites is much more sensitive to the variation of applied normal load than the sliding speed variation.

Tab. 2.4: Symptoms and appearance of different types of wear.

Types of wear	Symptoms	Appearance of the worn-out surface
Abrasive	Presence of clean furrows cut out by abrasive particles	Grooves
Adhesive	Metal transfer is the prime symptom	Seizure, cratering, rough and torn-out surfaces
Erosion	Presence of abrasives in the fast moving fluid and short abrasion furrows	Waves and troughs
Corrosion	Presence of metal corrosion products	Rough pits or depressions
Fatigue	Presence of surface or subsurface cracks accompanied by pits and spalls	Sharp and angular edges around pits
Impacts	Surface fatigue, small submicron particles or formation of spalls	Fragmentation, peeling and pitting
Delamination	Presence of subsurface cracks parallel to the surface with semi-dislodged or loose flakes	Loose, long and thin sheet-like particles
Fretting	Production of voluminous amount of loose debris	Roughening, seizure and development of oxide ridges
Electric attack	Presence of microcraters or a track with evidence of smooth molten metal	Smooth holes

Fig. 2.17 (a)–(c) show the SEM micrographs of worn surfaces of PO (parallel orientation), APO (anti-parallel orientation) and NO (normal orientation) samples respectively (refer to Fig. 2.18). In the PO sample, fiber orientation supports the flow of asperities. First the fiber's surface and cells are delaminated by these hard asperities, bent, and then the removal of the debris occurred from the composite surface. Whereas the remaining attached fiber became fibrillated, this is clearly visible in Fig. 2.17 (a). In this case delamination of fiber's cell and microplowing mechanism dominated in the wear process. Fig. 2.17 (b) shows the worn surface of the APO sample where the abrasive particles have slid perpendicular to the fiber alignment. Due to this transverse motion, the abrasives cut the fibers and bend them in the direction of sliding motion. The wear process in this case is mainly due to the microcutting of the fiber. In case of the APO sample, cells of fiber are oriented normal to the sliding direction. This fiber geometry resists the flow of asperities and removal of debris. Fig. 2.17 (c) shows the unworn surface of the NO sample, which illustrates the fibrillated cross section without fiber pullout. In this case microcutting of the fiber cross section is mainly responsible for formation of fine wear debris.

(a) (b) (c)

Fig. 2.17: SEM photographs of worn surface of different composites at 10 N load; (a) worn surface of PO-type sample, (b) worn surface of APO-type sample, (c) worn surface of NO-type sample [23].

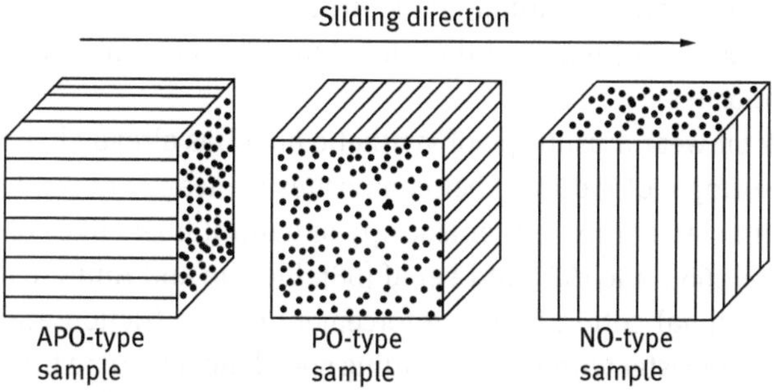

Fig. 2.18: Lantana camara fiber reinforced epoxy composite with three different fiber orientations with respect to sliding direction (APO, PO, and NO) [23].

2.6.5 Tribological anisotropy in natural fiber composites

Unidirectional, continuous fiber reinforced polymer composites exhibit significant tribological anisotropy due to their heterogeneity. As described in the literature [58], fiber orientation with respect to sliding direction is one of the most important parameters affecting properties of composites including friction and wear behavior of fiber reinforced polymer composite. It is also said that properties of natural fiber composites are influenced by fiber loading, dispersion, orientation, and fiber to matrix interface [59]. Natural fibers such as sisal and jute are naturally occurring composites containing cellulose fibrils embedded in lignin matrix. These cellulosic fibrils are aligned along the length of the fiber irrespective of its origin. Such an alignment leads to maximum tensile strength and provides rigidity in that direction of the fiber. Experimental investigation has shown that the largest wear resistance in fiber reinforced polymer composites occurred when the sliding was normal to the fiber orientation, while the lowest wear resistance occurred when the fiber orientation was in the transverse direction. Experiments have also shown that the coefficient of friction and the wear in fiber reinforced polymer composites depend on several factors including the material combination, the fiber orientation, and the surface roughness.

Cirino et al. [60] studied the dry abrasive wear behavior of continuous aramid fiber reinforced epoxy composite and found that among three orientations of aramid fiber in epoxy matrix, i.e., normal, parallel and anti-parallel directions, normal orientation produces optimum wear resistance. Shim et al. [61] reported the effect of fiber orientation on friction and wear properties of graphite fiber composites and discovered that the differences in friction and wear behavior of specimens with different fiber orientation are mainly due to the anisotropic properties caused by the microstructure of fiber orientation in the matrix. Lhymn [62] investigated the tribological properties of unidirectional polyphenylene sulfide-carbon fiber laminate composites and reports that fibers that are oriented normal to sliding surface exhibit better wear resistance. He also attempted to qualitatively explain the effect of fiber orientation in terms of the difference in the interlaminar shear strength and the fracture strain of the three principal fiber orientations. Sung et al. [58] reported the same type of result while they worked on Kevlar-epoxy composites.

Although extensive work on wear anisotropy of synthetic fiber has already been done, the work on wear anisotropic of natural fiber composites is meager. Recently some attempt has been taken to study the wear anisotropy of natural fibers like cotton [63], bamboo [64, 65], sisal [66], jute [67], and kenaf [68]. Eleiche and Amin [63] reported the effect of unidirectional cotton fiber reinforcement on the friction and sliding wear characteristics of polyester with varying sliding speed, fiber volume fraction, and fiber orientation. Chand et al. [64, 65] studied the wear behavior of bamboo in different orientations such as NO, PO and APO, and observed that in bamboo the wear rate follows the trend $W_{NO} < W_{APO} < W_{PO}$. Similar behavior is reported by them while working with sisal [66] and jute fibers [67]. Tong et al. [69] reported the three-body abrasive wear (low stress) results of bamboo against a free abrasive consisting of quartz sand and bentonite on a rotary-disk type abrasive wear tester. Chin et al. [68] investigated wear and friction performance of kenaf fiber reinforced epoxy composite in three different fiber orientations with respect to the sliding direction and reported that the composite exhibited better wear performance in NO compared to PO and APO orientations.

Variation of wear rate with applied load for different directionally oriented (PO, APO and NO) Lantana camara fiber reinforced epoxy composites have been studied [23]. It is clear from Fig. 2.19 that wear rate of the composite increases with the increase in normal load for different directionally oriented fibers. However, the magnitude of the wear rate is not the same. Minimum wear rate has been observed for NO sample, whereas PO sample exhibits maximum wear rate. The wear rate follows the trend; $W_{NO} < W_{APO} < W_{PO}$ which indicate anisotropic wear behavior. This behavior is similar to that of other fiber composites [64–67]. In the case of NO-type sample, the long fibers are well embedded in the matrix and only the cross sections of the vascular bundles come in contact with abrasive particle which oppose the movement of the abrading particles, as a result minimum wear occurred. There is a possibility of maximum real contact area with fibers in the sliding direction in the case of the PO-type sample, which leads to maximum wear in comparison to APO-type and NO-type sample. In this case, the abrasion in the composite might have taken place due to the removal of a complete layer of fiber, microcutting of the cell, delamination of fibers leading to microcutting and breaking of resin. This in turn leads to formation of debris. In APO-type sample, the exposed area of fiber is lesser in comparison to PO-type but higher than that in NO-type. The removal of complete fibers is restricted due to phase discontinuity, i.e., because of the presence of matrix phase present between the fibers. Chand et al. [65] report the same type of results when they studied the abrasive wear behavior of bamboo.

The wear anisotropy of unidirectional Lantana camara fiber reinforced epoxy composites depends on load and abrasive grit size. With increasing particle size (lower grit number), wear rate of composites increases. Again in the load range of 1–7 N, the value of the anisotropic coefficient increases. However, post 7 N, the coefficient decreases again. In the PO-type sample, the abrasion taking place is due to microplowing and delamination of fiber's cell, whereas in APO-type and NO-type samples, microcutting is found to be responsible for the wear process.

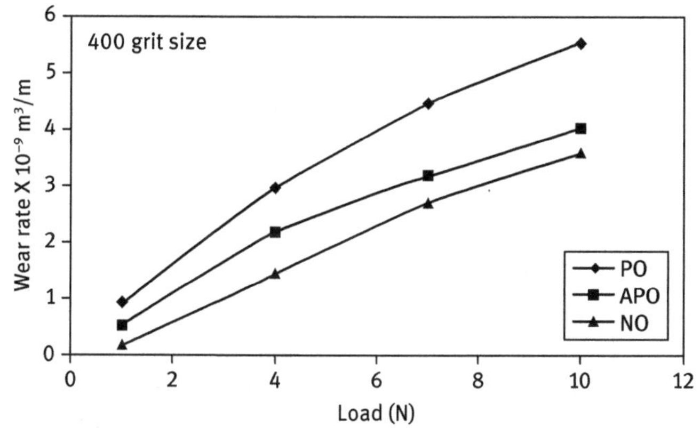

Fig. 2.19: Plot between wear rates versus applied load for different oriented composites at 27 m sliding distance and 400 grit size abrasive paper [23].

Fig. 2.20 shows that the increase in abrasive grit size from 400 to 100 increases the weight loss of the above three differently oriented Lantana camara fiber epoxy composites [23]. The wear rate is primarily dependent on the depth and width of the groove made by the abrasives. Using coarser abrasives, the depth of penetration of the abrasive particles is high, therefore material removal from the specimen surface is increased. If the applied load is fixed, then the effective stress on individual abrasive particles increases with coarser abrasive particles, as the load is shared by a lower number of abrasive particles. When the abrasive particles are finer in size, they make only elastic contact with the test specimen surface, as the effective stress in individual

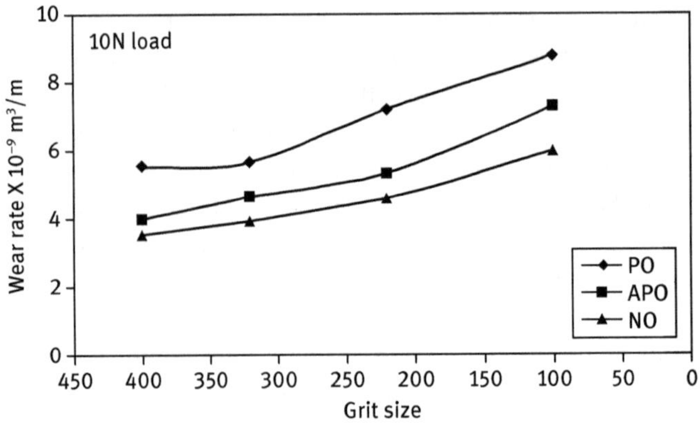

Fig. 2.20: Plot between wear rate vs grit size for PO, APO, and NO samples for applied load of 10 N, sliding distance: 27 m [23].

abrasive particles is less. However, at higher load regime, the effective stress on each individual abrasive particle reaches to a level where the abrasives make plastic contact with the specimen surface causing more surface damage even at finer abrasive size.

2.7 Closure

Increased technical innovation, identification of new applications, continuing political and environmental pressure and government investments in new methods for fiber harvesting and processing are leading to projections of continued growth in the use of natural fiber-based composites. Now, appreciating the importance of tribology in minimizing energy consumption, extending product life, and protecting the environment is quite relevant nowadays. The easy availability of natural fibers and manufacturing of their composites have motivated researchers worldwide to try locally available inexpensive fibers and to study their feasibility for reinforcement purposes and to what extent they satisfy the required specifications of a good reinforced polymer composite for tribological applications. It has been found that the application of natural fibers has quite a positive effect on the tribological performance of the composites. However, natural fibers also suffer from some disadvantages like water absorption, lower adhesion between fiber and matrix and above all geographical variations. To offset these limitations, development of hybrid composites has started taking place. Besides, continuous advancement in processing techniques has made the natural fiber-based composites suitable for some industries, especially the automobile sector. However, with continuous awareness and technological development, the acceptance and applicability of green composites will surely experience newer heights.

References

[1] Mahapatra SS, Patnaik A. Study on mechanical and erosion wear behavior of hybrid composites using Taguchi experimental design. Mater Design. 2009;30(8):2791–2801.

[2] Sahoo P. Engineering tribology. New Delhi: Prentice Hall of India; 2005.

[3] Raghavendra G, Ojha S, Acharya SK, Pal SK. Mechanical properties of woven jute-glass hybrid-reinforced epoxy composite. J Compos Mater. 2013;48(28):3445–3455.

[4] Nirmal U, Hashim J, Megat Ahmad MMH. A review on tribological performance of natural fibre polymeric composites. Tribol Int. 2015;83:77–104.

[5] Mohammed L, Ansari MNM, Pua G, Jawaid M, Saiful Islam M. A review on natural fiber reinforced polymer composite and its applications. Int J Polym Sci. 2015;2015(ID 243947):1–15.

[6] Raghavendra G. Mechanical and tribological behavior of nanofiller reinforced polymer nanocomposite. PhD thesis. National Institute of Technology, Rourkela, India; 2014.

[7] Thakur VK, Thakur MK. Processing and characterization of natural cellulose fibers/thermoset polymer composites. Carbohyd Polym. 2014;109;102–117.

[8] Sanjay MR, Arpitha GR, Laxmana Naik L, Gopalakrishna K, Yogesha B. Applications of natural fibers and its composites: An overview. Natural Resources. 2016;7:108–114.

[9] Koronis G, Silva A, Fontul M. Green composites: A review of adequate materials for automotive applications. Compos Part B Eng. 2013;44(1):120–127.

[10] John MJ, Thomas S. Biofibres and biocomposites. Carbohyd Polym. 2008;71(3):343–364.

[11] Davoodi MM, Sapuan SM, Ahmad D, Ali A, Khalina A, Jonoobi M. Mechanical properties of hybrid kenaf/glass reinforced epoxy composite for passenger car bumper beam. Mater Design. 2010;31(10):4927–4932.

[12] Samal SK, Mohanty S, Nayak SK. Banana/glass fiber-reinforced polypropylene hybrid composites: fabrication and performance evaluation. Polym Plast Technol. 2009;48(4):397–414.

[13] Lu JZ, Qinglin W, McNabb HS Jr. Chemical coupling in wood fiber and polymer composites: a review of coupling agents and treatments. Wood and Fiber Sci. 2000;32(1):88–104.

[14] Mohanty AK, Misra M, Drzal LT. Surface modifications of natural fibers and performance of the resulting biocomposites: An overview. Compos Interface. 2001;8(5):313–343.

[15] Agrawal R, Saxena NS, Sharma KB, Thomas S, Sreekala MS. Activation energy and crystallization kinetics of untreated and treated oil palm fibre reinforced phenol formaldehyde composites. Mater Sci Eng. 2000;277(1–2):77–82.

[16] Jähn A, Schröder MW, Füting M, Schenzel K, Diepenbrock W. Characterization of alkali treated flax fibres by means of FT Raman spectroscopy and environmental scanning electron microscopy. Spectrochim Acta A. 2002;58(10):2271–2279.

[17] Valadez-Gonzalez A, Cervantes-Uc JM, Olayo R, Herrera-Franco PJ. Effect of fiber surface treatment on the fiber-matrix bond strength of natural fiber reinforced composites. Compos Part B Eng. 1999;30(3):309–320.

[18] Nair KC, Diwan SM, Thomas S. Tensile properties of short sisal fiber reinforced polystyrene composites. J Appl Polym Sci. 1996;60(9):1483–1497.

[19] Paul SA, Joseph K, Mathew GDG, Pothen LA, Thomas S. Influence of polarity parameters on the mechanical properties of composites from polypropylene fiber and short banana fiber. Compos Part A Appl S. 2010;41(10):1380–1387.

[20] Li X, Tabil LG, Panigrahi S, Crerar WJ. Flax fiber-reinforced composites and the effect of chemical treatments on their properties. North Central Regional Annual Meeting ASAE/CSAE Conference; 2004.

[21] Srinivasan VS, Rajendra Boopathy S, Sangeetha D, Vijaya Ramnath B. Evaluation of mechanical and thermal properties of banana flax based natural fibre composite. Mater Design. 2014;60:620–627.

[22] Shalwan A, Yousif BF. In State of Art: Mechanical and tribological behaviour of polymeric composites based on natural fibres. Mater Design. 2013;48:14–24.
[23] Deo C. Preparation and characterization of polymer matrix composite using natural fiber Lantana-camara. PhD thesis, National Institute of Technology, Rourkela, India; 2010.
[24] Rahman M, Huque M, Islam M, Hasan M. Improvement of physico-mechanical properties of jute fiber reinforced polypropylene composites by post-treatment. Compos Part A Appl S. 2008;39(11):1739–1747.
[25] John K, Naidu SV. Tensile properties of unsaturated polyester-based sisal fiber-glass fiber hybrid composites. J of Reinforced Plast and Composites. 2004;23(17):1815–1819.
[26] Mishra S, Mohanty AK, Drzal LT, et al. Studies on mechanical performance of biofibre/glass reinforced polyester hybrid composites. Compos Sci Technol. 2003;63(10):1377–1385.
[27] Almeida JHS Jr., Amico SC, Botelho EC, Amado FDR. Hybridization effect on the mechanical properties of curaua/glass fiber composites. Compos Part B Eng. 2013;55:492–497.
[28] Rong MZ, Zhang MQ, Liu Y, Yang GC, Zeng HM. The effect of fiber treatment on the mechanical properties of unidirectional sisal-reinforced epoxy composites. Compos Sci Technol. 2001;61(10):1437–1447.
[29] Pavithran C, Mukherjee PS, Brahmakumar M, Damodaran AD. Impact properties of sisal-glass hybrid laminates. J Mater Sci 1991;26(2):455–459.
[30] Pavithran C, Mukherjee PS, Brahmakumar M. Coir-glass intermingled fibre hybrid composites. J of Reinforced Plast and Composites. 1991;10(1):91–101.
[31] Kalaprasad G, Thomas S, Pavithran C, Neelakantan NR, Balakrishnan S. Hybrid effect in the mechanical properties of short sisal/glass hybrid fiber reinforced low density polyethylene composites. J of Reinforced Plast and Composites. 1996;15(1):48–73.
[32] Saba N, Tahir MP, Jawaid M. A review on potentiality of nano filler/natural fiber filled polymer hybrid composites. Polymers. 2014;6(8).
[33] Borba PM, Tedesco A, Lenz DM. Effect of reinforcement nanoparticles addition on mechanical properties of SBS/curauá fiber composites. Mat Res. 2014;17:412–419.
[34] Jawaid M, Abdul Khalil HPS. Cellulosic/synthetic fibre reinforced polymer hybrid composites: A review. Carbohyd Polym. 2011;86(1):1–18.
[35] Sathishkumar TP, Naveen J, Satheeshkumar S. Hybrid fiber reinforced polymer composites – a review. J of Reinforced Plast and Composites. 2014;33(5):454–471.
[36] Fu SY, Feng XQ, Lauke B, Mai YW. Effects of particle size, particle/matrix interface adhesion and particle loading on mechanical properties of particulate-polymer composites. Compos Part B: Eng. 2008;39(6):933–961.
[37] Camargo PHC, Satyanarayana KG, Wypych F. Nanocomposites: synthesis, structure, properties and new application opportunities. Mat Res. 2009;12:1–39.
[38] Sahoo P, Das SK, Davim JP. Surface finish coatings. In: Saleem Hashmi, editor. Comprehensive materials finishing. 1st ed. Elsevier; 2016. p. 38–55.
[39] Bajpai PK, Singh I, Madaan J. Tribological behavior of natural fiber reinforced PLA composites. Wear. 2013;297(1–2):829–840.
[40] El-Tayeb NSM. A study on the potential of sugarcane fibers/polyester composite for tribological applications. Wear. 2008;265(1–2):223–235.
[41] Yousif BF, El-Tayeb NSM. Adhesive wear performance of T-OPRP and UT-OPRP composites. Tribol Lett. 2008;32(3):199–208.
[42] Eyre TS. Wear characteristics of metals. Tribol Int. 1976;9(5):203–212.
[43] Deo C, Acharya SK. Effects of fiber content on abrasive wear of Lantana camara fibre reinforced polymer matric composite. Indian J Eng Mat S. 2010;17:219–223.
[44] Deo C, Acharya SK. Effects of load and sliding velocity on abrasive wear of Lantana camara fibre-reinforced polymer matrix composite. P I Mech Eng J-J Eng. 2010;224(5):491–496.

[45] Singh Gill N, Yousif BF. Wear and frictional performance of betelnut fibre-reinforced polyester composite. P I Mech Eng J-J Eng. 2008;223(2):183–194.
[46] Pool KV, Dharan CKH, Finnie I. Erosive wear of composite materials. Wear. 1986;107(1):1–12.
[47] Bhushan B. Principles and applications of tribology. Second edition. Wiley; 2013.
[48] Harsha AP, Tewari US, Venkatraman B. Solid particle erosion behaviour of various polyaryl-etherketone composites. Wear. 2003;254(7–8):693–712.
[49] Mishra P, Acharya SK. Anisotropy abrasive wear behavior of bagasse fiber reinforced polymer composite. Int J Eng Sci Technol. 2010;2(11):104–112.
[50] Deo C, Acharya SK. Solid particle erosion of lantana camara fiber-reinforced polymer matrix composite. Polym Plast Technol. 2009;48(10):1084–1087.
[51] Gupta A, Kumar A, Patnaik A, Biswas S. Effect of different parameters on mechanical and erosion wear behavior of bamboo fiber reinforced epoxy composites. Int J Polym Sci. 2011;(592906):1–10.
[52] Mohanty JR, Das SN, Das CD, Mahanta TK, Sataya B. Solid particle erosion of date palm leaf fiber reinforced polyvinyl alcohol composites. Advances in Tribology. 2014;2014(Article ID 293953):1–8.
[53] Arjula S, Harsha AP. Study of erosion efficiency of polymers and polymer composites. Polym Test. 2006;25(2):188–196.
[54] Biswas S, Satapathy A. A comparative study on erosion characteristics of red mud filled bamboo-epoxy and glass-epoxy composites. Mater Design. 2010;31(4):1752–1767.
[55] Raghavendra G, Ojha S, Acharya SK, Pal SK. A comparative analysis of woven jute/glass hybrid polymer composite with and without reinforcing of fly ash particles. Polym Compos. 2016;37(3):658–665.
[56] Barkoula NM, Karger-Kocsis J. Effects of fibre content and relative fibre-orientation on the solid particle erosion of GF/PP composites. Wear. 2002;252(1–2):80–87.
[57] Dasari A, Yu ZZ, Mai YW. Fundamental aspects and recent progress on wear/scratch damage in polymer nanocomposites. Mat Sci Eng R. 2009;63(2):31–80.
[58] Sung N-H, Suh NP. Effect of fiber orientation on friction and wear of fiber reinforced polymeric composites. Wear. 1979;53(1):129–141.
[59] Joseph K, Varghese S, Kalaprasad G, et al. Influence of interfacial adhesion on the mechanical properties and fracture behaviour of short sisal fibre reinforced polymer composites. Eur Polym J. 1996;32(10):1243–1250.
[60] Cirino M, Friedrich K, Pipes RB. The effect of fiber orientation on the abrasive wear behavior of polymer composite materials. Wear. 1988;121(2):127–141.
[61] Shim HH, Kwon OK, Youn JR. Effects of fiber orientation and humidity on friction and wear properties of graphite fiber composites. Wear. 1992;157(1):141–149.
[62] Lhymn C. Tribological properties of unidirectional polyphenylene sulfide-carbon fiber laminate composites. Wear. 1987;117(2):147–159.
[63] Eleiche AM, Amin GM. The effect of unidirectional cotton fibre reinforcement on the friction and wear characteristics of polyester. Wear. 1986;112(1):67–78.
[64] Chand N, Dwivedi UK, Acharya SK. Anisotropic abrasive wear behaviour of bamboo (Dentrocalamus strictus). Wear. 2007;262(9–10):1031–1037.
[65] Chand N, Dwivedi UK. High stress abrasive wear study on bamboo. J Mater Process Technol. 2007;183(2–3):155–159.
[66] Chand N, Dwivedi UK. Influence of fiber orientation on high stress wear behavior of sisal fiber-reinforced epoxy composites. Polym Compos. 2007;28(4):437–441.
[67] Dwivedi UK, Chand N. Influence of fibre orientation on friction and sliding wear behaviour of jute fibre reinforced polyester composite. Appl Compos Mat. 2009;16(2):93–100.

[68] Chin CW, Yousif BF. Potential of kenaf fibres as reinforcement for tribological applications. Wear. 2009;267(9–10):1550–1557.
[69] Tong J, Ren L, Li J, Chen B. Abrasive wear behaviour of bamboo. Tribol Int. 1995;28(5):323–327.

S. Manoharan, V. Krishnaraj, R. Vijay, D. Lenin Singaravelu, and B. Suresha

3 Development and characterization of novel fiber reinforced hybrid friction composites

3.1 Introduction

Fiber reinforced polymer matrix composites (PMCs) are primarily finding their applications in automotive and railway systems as brake friction composites, where friction and wear are critical issues [1]. Fiber reinforced friction composites (brake pads) employed in a brake system often comprise an organic polymer matrix (phenol formaldehyde resin) inside which a number of reinforcing short fibers (glass, carbon, mineral, synthetic, and aramid fibers), frictional additives (molykote, copper, silicon carbide, and alumina) and fillers (barium sulfate and cashew friction dust) are distributed. In the friction composites, the fibers play a very vital role in maintaining the higher and stable coefficient of friction (μ), less brake noise and wear than the asbestos lining, wear minimization and friction optimization under operating variables such as braking force, sliding speed, braking duration and braking temperature [2]. The non-asbestos organic (NAO) materials have been found to be less ductile and to display lower wear rates when compared to other organic friction composites. The design of formulation of friction composites is a well-known problem in multicriteria optimization. This is because it involves handling four prime classes of constituents and also reaching a suitable and desired level of performance attributes. In fact, the formulation of a composite for a brake system should be designed to satisfy a number of requirements for the properties, such as durability, good tribological characteristics and low noise at a wide range of operating conditions [3].

Adhesive and abrasive wear are the major noticeable areas in various polymer-based friction composites [4]. The abrasive wear of engineering and automotive components caused by abrasive particles is a major industrial problem. The performance of a friction composite is highly related to the formation of a heterogeneous friction film (third-body layer) at the friction interface [5]. While braking, heat is generated at the friction interface and then allowed to cool down. This causes a shear-induced thermomechanical stressed state in the friction composite. Due to this state, the heterogeneous friction film gets reduced in deformation and becomes hardened, imparting a brittleness effect on the friction film. Hence, the brittleness effect causes disintegration of the heterogeneous friction film and generates hard wear debris particles between the friction surfaces. These wear debris particles are entrapped in the mating zone of the friction composite and the disc and act as hard abrasives in the form of third bodies, resulting in three-body abrasion (TBA). This is a major automotive industrial setback [3, 6]. Abrasive wear by sand particles is an important factor causing excessive wear in machines operating in sandy environments and is another industrial problem.

https://doi.org/10.1515/9783110435788-004

This abrasion occurs due to the inflow and entrapment of sand particles at the friction interface [7]. Hence, to prevent such machines from failure, it is important to understand how high stress three-abrasive wear occurs in the friction composites during abrasion.

Though many researchers have explored the basic mechanism of abrasive wear in PMCs, a focus on three-body abrasion of friction composites has been a long-felt need. Chand et al. [8] studied the three-body abrasion of short E-glass fiber reinforced PMCs using silica sand size of 100 to 200 µm and 200 to 300 µm as abrasives and found that the abrasive wear of the composite depended on test parameters such as applied load, sliding speed, weight fraction of fiber and size of abrasive particle. Cenna et al. [9] studied abrasion resistance of vinyl ester-based PMCs using hard ignimbrite and soft coal bulk solids as abrasives and suggested that the coarse ignimbrite abrasive particles are hard and retain their shape and angularity for a longer duration of time, which effectively removes the material at a faster rate when compared to soft coal particles. Öztürk et al. [10] studied the tribological behavior of basalt and ceramic fiber reinforced hybrid friction material. Ceramic fiber was maintained constant (10 vol.%) and basalt fiber was varied from 0 to 40 vol.% by compensating $BaSO_4$ content and it was concluded that the addition of basalt fiber increases the friction coefficient and the wear resistance of hybrid friction composites. More than 30 vol.% of basalt fiber increases the specific wear rate of friction composites.

During braking, the normal operating temperature of the rotor disc can reach up to 200–250 °C in the case of a passenger car and in the disc pads a high temperature is evident up to 400 °C. While braking, the temporary reduction in friction coefficient due to the loss of friction between the braking surfaces as a consequence of increased interface temperature is termed fade. However, after a temporary exposure to high temperature and braking-induced thermal elevations, the extent of restoration of the original magnitude of μ on the release of brakes and subsequent cooling down of the pad is termed recovery [11]. Thus to avoid these thermal-induced performance complications, it is necessary to reduce thermoelastic instability (TEI) at the dynamically mating interfaces. Satapathy and Bijwe [11] developed friction composites based on varying a combination of organic (aramid and PAN) and rock (lapinus and basalt) fibers and concluded that a higher content of shot in the basalt fiber led to a higher μ, resulting in a higher degree of wear when compared to the lapinus fiber. The aramid fiber also led to a higher μ and a higher wear resistance when compared to the PAN fiber.

Most experimental works detailed the effect of one factor by maintaining all the other factors as constants. But, the combined effect of interacting factors that influence the experimentation has not been studied. Using the Taguchi method, it is easy to investigate the influence of each component in problematical friction composites. In fact, this technique works out for optimizing only single performance characteristics. To overcome this problem, some authors have coupled the Taguchi method with the Grey Relational Analysis (GRA) to solve multiple performance characteristics and to optimize the tribological parameters with multiple responses. Subbaya et al. [12] conducted three-body abrasive wear studies on carbon fabric reinforced epoxy (C-E) composites by adding graphite particles as secondary reinforcement. The specific wear rate, wear coefficient and hardness were considered as response factors to be optimized. Although the optimization process involves various aspects, it is essential to optimize the tribological behaviors of friction composites in order to identify best combinational parameters. By reviewing the above studies, it was decided to optimize the three-body abrasive wear behavior of hybrid friction composites using various controlling parameters such as abrasive particle size (A), applied load (L), wt% of fiber content (F) and abrading distance (D). A GRA-Taguchi approach was employed to predict the optimal parameter combinations for the multiple response characteristics of friction composites.

3.1.1 Research gap based on literature inference

Fibers are mainly used for reinforcement in place of conventional asbestos, which causes carcinogenic problems. Several researchers have reported the influence of fibers such as metal, mineral, carbon, glass, aramid and organic fibers individually or in combination with other groups on tribological behaviors of friction composites. However, not many studies about the effect of a basalt and recycled aramid fiber combination on thermal, mechanical and tribological behavior of hybrid friction composites under various operating conditions are available. Although three-body abrasion is common for PMCs it has not been explored particularly well in friction material composites and is still not well understood. The grounds for three-body abrasive wear in friction composites have been explained in this section. Hence, considering the above mentioned facts, it was decided to study the effect of basalt and recycled aramid fiber reinforcement on the thermal, mechanical and tribological behavior of hybrid friction composites.

3.1.2 Objectives of the present study

The current research work focuses on the following objectives:

- To fabricate hybrid friction composites based on five different wt% of fibers using hot compression molding method.
- To evaluate the effect of fiber reinforcement on physical (density and water swelling), chemical (acetone extraction), thermal (heat swelling and loss of weight on ignition), mechanical (hardness, tensile, compressive, flexural and impact, and thermogravimetric analysis) properties of friction composites.
- To understand the effect of fiber reinforcement on three-body abrasive wear behavior of friction composites. To find out the optimum factor combination for

multiple responses using Grey–Taguchi method by optimizing the specific wear rate, wear coefficient and hardness in three-body abrasive wear conditions.
- To study the effect of fiber reinforcement on fade and recovery behaviors of hybrid friction composites. To study the worn surface morphologies of hybrid friction composites

3.2 Materials and method

A flow chart showing the experimental procedure related to friction composites is shown in Fig. 3.1. The experimental details related to hybrid fiber reinforced friction composites in accordance with ASTM and industrial standards are also discussed in this section.

3.2.1 Materials

In the present work, the ingredients which have been chosen for developing the friction composites are listed in Tab. 3.1 and SEM images of fibers are shown in Fig. 3.2 (a) and (b).

3.2.1.1 Basalt fiber

Basalt fiber is a hard and dense alumino-silicate mineral fiber composed of many oxides and igneous rock. It is solidified volcanic lava of natural origin with high strength and modulus, hardness, heat insulating properties and good fiber/matrix interfacial bonding and abrasion resistance that could be related to the degree of crystallinity as

Tab. 3.1: Design of formulation of hybrid friction composites (ingredients are in wt%).

S. No.	Composite designation	Reinforcements 1. Basalt fiber 2. Recycled aramid fiber			Binders 1. Phenol formaldehyde 2. Cashew nut shell liquid 3. Plaster of Paris 4. Carbon black					Fillers 1. Molykote 2. Cu powder 3. Iron Powder 4. SiC powder 5. Al_2O_3 powder 6. Cashew friction dust							$BaSO_4$	Total
1	C1	20	5	**25**	14	9	4	3	**30**	11	7	7	4	5	11	**45**	**0**	**100**
2	C2	15	5	**20**	14	9	4	3	**30**	11	7	7	4	5	11	**45**	**5**	**100**
3	C3	10	5	**15**	14	9	4	3	**30**	11	7	7	4	5	11	**45**	**10**	**100**
4	C4	5	5	**10**	14	9	4	3	**30**	11	7	7	4	5	11	**45**	**15**	**100**
5	C5	5	0	**5**	14	9	4	3	**30**	11	7	7	4	5	11	**45**	**20**	**100**

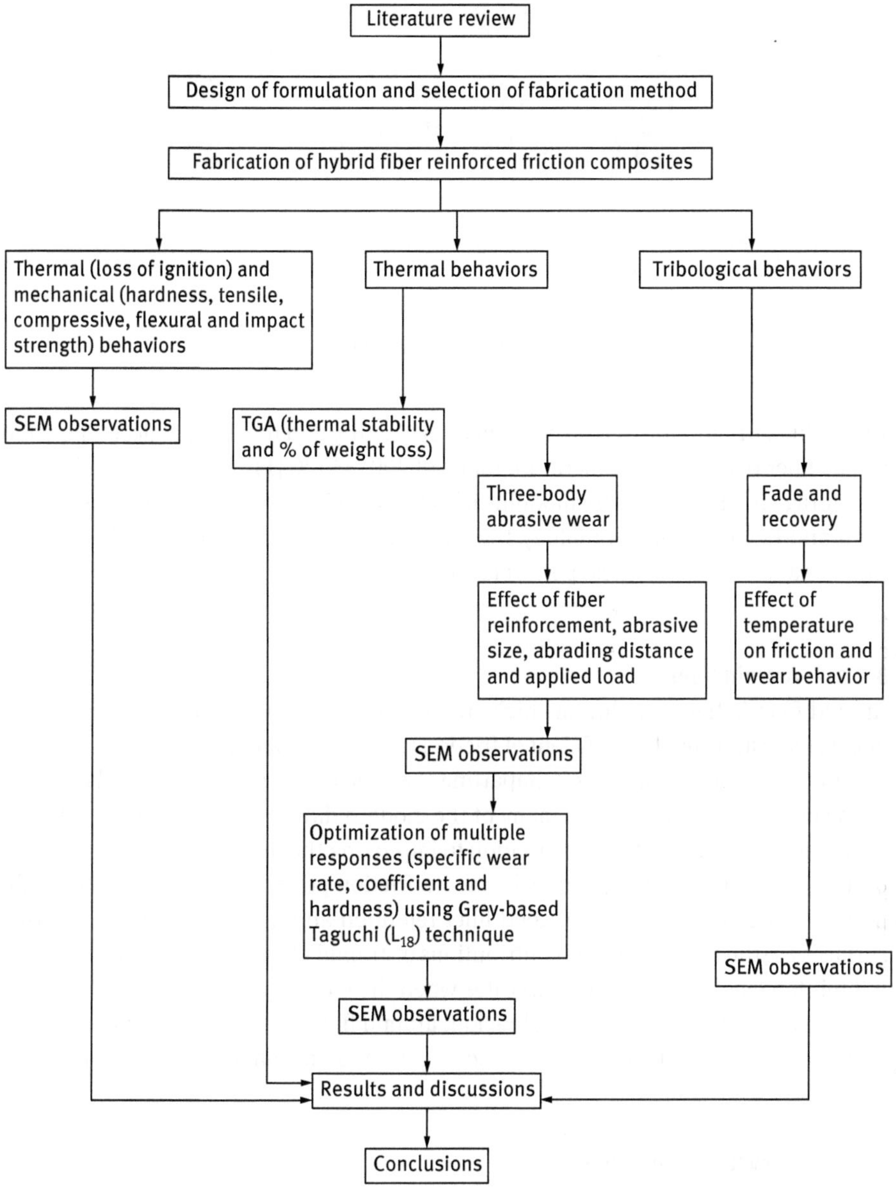

Fig. 3.1: Experimental flow chart of hybrid friction composites.

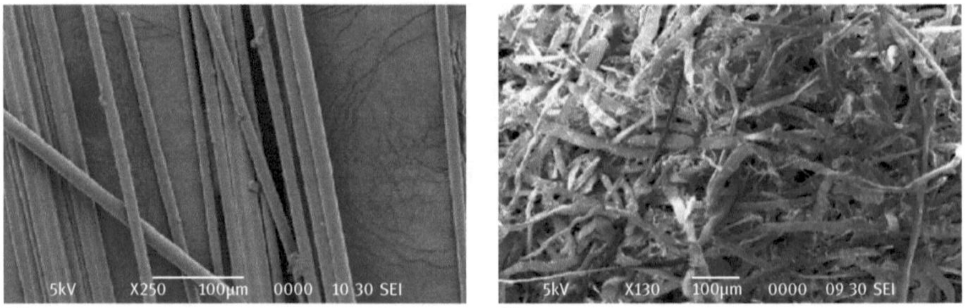

(a) (b)

Fig. 3.2: SEM images of (a) basalt fiber (b) recycled aramid fiber.

well as fiber orientation. It has better thermal stability and the thermal decomposition of this fiber occurs in a temperature range of 200–350 °C [13]. Basalt fiber influences the storage modulus and loss modulus in PMCs significantly [14]. The basalt fiber has a tail-and-head or shot morphology, the "shot" being the harder part of the basalt fiber that enhances its tribological performance [11].

3.2.1.2 Aramid fiber

Aramid fiber is light weight, has high strength and modulus, high thermal stability, and a stable and steady coefficient of friction (μ). Aramid fiber is a synergistic in combination with rock wool fibers in imparting low fade characteristics. Aramid fiber has been successfully employed in many of the modern day non-asbestos organic (NAO) formulations because of its contribution to improving the operational life expectancy of the brakes, imparting low squeal, smooth friction performance, and better toughness. Superior wear resistance, good filler retention and nonabrasiveness to mating surfaces are the some of the key contributions of aramid fiber, which makes it an essential reinforcement in friction composites when the composites are subjected to abrasive wear conditions [15]. During abrasion, aramid fiber improves the wear resistance of the composite due to its characteristic properties of fibrillation and elongation [16].

3.2.1.3 Recycled aramid fiber

In general, to recycle waste in order to prevent dumping of solid wastes, the fibers are recycled from the used materials. Also, the recycled aramid fibers possess the same properties as the virgin aramid fibers and provide these in a with cost effective manner.

3.2.2 Processing

The details of the fabrication process of hybrid friction composites containing thirteen ingredients are tabulated in Tab. 3.2. The fabrication process consists of a series of operations including mixing, hydraulic pressing, curing, post-curing and finishing as shown in Fig. 3.3 (a) and (b). In order to obtain the homogeneous mix, the sequence and time of mixing were maintained properly. The proper proportions of the various ingredients were blended in a laboratory mixer. Curing refers to the toughening or hardening of a composite made from several different materials. This is achieved using chemical additives or heat or with the use of ultraviolet radiation. In the present study, the post-curing process helps in giving a stronger, more reliable product which can be produced more efficiently and cost effectively. Post-curing operation was done to cure the residual resin. Also it increases the amount of crosslinking of polymer to achieve better chemical and mechanical properties and heat resistance [17]. Finally the composite block of size 200 mm × 200 mm × 6 mm was removed from the mold and the required size of sample was prepared according to experimental work. Test samples for various experiments were prepared from the friction composites using abrasive cut-off machine according to ASTM standard.

Tab. 3.2: Details of fabrication conditions of hybrid friction composites.

Procedure	Fabrication conditions
Sequence of mixing	Lodigee type shear mixer – Chopper speed: 2800 rpm – Rotor speed: 1440 rpm – Total duration: 20 min – Sequence: (a) fibers (basalt and recycled aramid) are mixed for 5 min, (b) filler particles were added and further mixed about 10 min, (c) phenol formaldehyde-based binder was added and final mix for about another 5 min.
Curing-compression molding process	– Temperature: 150 °C – Compression: 40 t – Curing time: 15 min
Post-curing	After curing, the composite plate was removed from the mold and post-cured at: – 120 °C for 60 min, – 150 °C for 60 min, – 170 °C for 120 min.

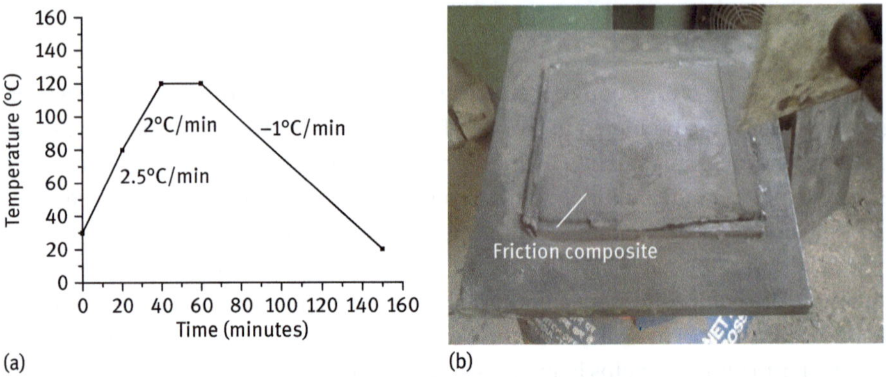

Fig. 3.3: (a) Curing cycle process, (b) hybrid friction composite with plate.

3.2.3 Experimentation

The details of the different experimentation work carried out on the hybrid friction composites under various standards are tabulated in Tab. 3.3.

Tab. 3.3: Experimentation details of hybrid friction composites.

S. No.	Name of experiment	Make	Operating conditions	Test standards
1	Density	(Mettler; Toledo) weighing balance	Archimedes principle	ASTM D-792
2	Water swelling		Temperature: ambient for a period of 30 min	IS 2742 Part 3
3	Acetone extraction	Soxhlet apparatus	Sample weight: 2 ± 0.2 g; Evaporation: about 10 min at about 80–85 °C	IS 2742 Part 3
4	Heat swelling		Temperature: 200 ± 3 °C at a time interval of 30–60 min	IS 2742 Part 3
5	Loss of weight on ignition at 800 °C	Muffle furnace	Sample weight: 1 g; Temperature: 800–850 °C	IS 2742 Part 3
6	Hardness test (Shore-D)	HT-6510D, Korean make	External load: 45 N, hardened steel rod with 1.4 mm diameter, 30° conical point, and 0.1 mm radius	ASTM D-2240

Tab. 3.3: (continued)

S. No.	Name of experiment	Make	Operating conditions	Test standards
7	Tensile test	Kalpak universal testing machine	Sample size: 165 × 13 × 6 mm^3; Load range: 10 kN; Test speed: 2 mm/min; Gauge length: 55 mm	ASTM D-638
8	Compressive test	Kalpak universal testing machine	Sample length: 25 mm; Diameter: 12.7 mm (½ inch) pin; Speed: 2 mm/min	ASTM D-695
9	Flexural test	Kalpak universal testing machine	Sample size: 130 × 13 × 6 mm^3; Span length: 64 mm; Load: 10 kN	ASTM D-790
10	Impact test	Izod pendulum Impact tester	Sample size: 65 × 13 × 6 mm^3; 45° notch	ASTM D-256
11	Thermogravimetric analysis/differential thermogravimetric (TG/DTG)	STA 449 F3 Jupiter	Temperature: room temperature to 800 °C; Heating rate: 10 K/min; Atmosphere: air; Sample weight: 2.5 mg	As per IUPAC Std
12	(a) Three-body abrasive wear test (routine test)	Rubber wheel abrasion test (RWAT) rig	Applied load: 40 N; Sample size: 70 × 27 × 6 mm^3; Abrasive size: 212 and 425 μm; Abrading distance: 250–1000 m increment of 250 m; Fiber content: 5, 10, 15, 20, and 25 wt%	ASTM G65
	(b) Three-body abrasive wear test (Taguchi L_{18} orthogonal array)	Rubber wheel abrasion test (RWAT) rig	Applied load: 30, 40, and 50 N; Abrasive size: 212 and 425 μm; Abrading distance: 1000, 1500, and 2000 m; Fiber content: 15, 20 and 25 wt%	ASTM G65
13	Fade and recovery test	Chase friction test machine	Sample size: 25 mm × 25 mm (625 mm^2) and 3 mm thick; Max. applied load: 667 N; Max. applied speed: 417 rpm; Operating temperature: 50–350 °C	SAEJ661/IS 2742 Part 4
14	Scanning Electron Microscopy	SEM; JEOL make 840 Model, Japan	Gold sputter coating	As per IUPAC Std

3.2.3.1 Evaluation of tribological behaviors and optimum process parameters

3.2.3.1.1 Three-body abrasive wear test (routine test)

A three-body abrasive wear test was conducted using dry sand rubber wheel abrasion test (RWAT) rig according to ASTM-G65 standard using silica sand (size of 212 and 425 µm) as abrasive, as shown in Fig. 3.4 (a) and (b). The experiments were carried out for constant static load of 40 N at a constant sliding velocity of 2.38 m/s. The abrading distances were varied in steps of 250 m, from 250 to 1000 m. For the second, (i.e., longer duration test involving, say, 500 m distance) the abrasion tests were carried out on the very same wear track where the first (i.e., 250 m) shorter runs were conducted [18]. The wear was measured by the loss in weight, which was then converted into wear volume using the measured density data. The specific wear rate (K_S) was calculated from the following equation:

$$K_S = \frac{\Delta V}{L \times D} \text{ m}^3/\text{Nm}. \tag{3.1}$$

In unlubricated wear, the value of wear coefficient (k) is the dimensionless quotient of the composite specimens [19]. The wear coefficient (k) was calculated from the following equation:

$$k = \frac{\Delta V H}{L \times D}, \tag{3.2}$$

where ΔV is volumetric wear, H is hardness of the wearing material, L is the applied normal load in Newton, and D is abrading distance in meters.

Fig. 3.4: SEM images of silica sand (a) 212 µm and (b) 425 µm.

3.2.3.1.2 Experimental design

The Taguchi technique is a method which conducts completely randomized simple experiments and the Taguchi parameter design can optimize the performance characteristics by setting the design parameters and by reducing the sensitivity of system performance to the source of variation [20, 21]. Grey relational analysis (GRA) measures and correlates the data difference between sequences and computes grey relational grade (GRG). To perform GRA, the experiment results are first normalized between zero and one, which is known as data preprocessing (grey relational generation). From the normalized data, the grey relational coefficient $\xi_i(k)$, which stands for correlation between desired and the actual experimental data, is calculated. By averaging grey relational coefficient (GRC) of each performance, the value of GRG which represents the overall evaluation of multiple performance characteristics is computed.

3.2.3.1.3 Data preprocessing

Data preprocessing is essential as the range and the unit in one data sequence may differ from others. In this process, the experimental data are normalized between zero and one and the original sequence is transferred into a comparable sequence. To obtain the optimal wear behavior, the "lower-the-better" characteristic has been used and it is normalized by using the following expression:

$$x_i^*(k) = \frac{\max x_i(k) - x_i(k)}{\max x_i(k) - \min x_i(k)}, \tag{3.3}$$

where $x_i^*(k)$ and $x_i(k)$ are sequences after data preprocessing and comparability sequence respectively, $k = 1, 2$ for specific wear rate and wear coefficient, $i = 1, 2, \ldots, 18$ for experiment number 1–18. The "higher-the-better" criterion can be expressed as

$$x_i^*(k) = \frac{x_i(k) - \min x_i(k)}{\max x_i(k) - \min x_i(k)}, \tag{3.4}$$

where $x_i^*(k)$ and $x_i(k)$ are sequences after data preprocessing and comparability sequence respectively, $k = 3$ for hardness, $i = 1, 2, \ldots, 18$ for experiment number 1–18.

$\Delta_{0i}(k)$ is the deviation sequence between reference sequence $x_0^*(k)$ and comparability sequence $x_i^*(k)$ and it is the difference of absolute value of $x_0^*(k)$ which is calculated using the following equation

$$\Delta_{0i}(k) = |x_0^*(k) - x_i^*(k)|. \tag{3.5}$$

3.2.3.1.4 Grey relational coefficient and grey relational grade

The grey relational coefficient stands for the correlation between desired and actual normalized experimental results and is given as follows:

$$\xi_i(k) = \frac{\Delta_{\min} + \zeta\,\Delta_{\max}}{\Delta_{0i}(k) + \zeta\,\Delta_{\max}} \tag{3.6}$$

$\xi_i(k)$ is the grey relational coefficient, $\Delta_{0i}(k)$ is the deviation sequence of reference sequence $x_0^*(k)$ and comparability sequence $x_i^*(k)$, ζ is distinguishing and identification coefficient. The purpose of GRC is to expand or to compress the range of the grey relational coefficient [22]. If all the parameters are given equal preference, ξ is taken as 0.5, which will weaken the effect of Δ_{max} when it is too big and defined in the range of $0 \leq \zeta \leq 1$. If all the parameters are given equal preference $\zeta = 0.5$ is used. After calculating GRC, the grey relational grade is computed by averaging the grey relational coefficient corresponding to each performance characteristic. The GRG-based multiple performance characteristics of the whole experiment can be evaluated by the following equation:

$$y_i = \frac{1}{n} \sum_{k=1}^{n} \xi_i(k). \tag{3.7}$$

y_i: grey relational grade,
$\xi_i(k)$: grey relational coefficient,
n: number of experiments.

3.2.3.1.5 Analysis of variance

Analysis of variance (ANOVA) is a statistical technique which reveals the level of significance of the main factor and its interactions on a particular response by comparing the mean square against errors at specific confidence levels. This is obtained by separating the total variability of the GRG, which is measured by the sum of the squared deviations from the total mean of GRG, into contributions by each influencing factor and the error. Percentage contribution is defined as the significant rate of parameters on a particular quality characteristic and indicates the power of the factor. This helps to investigate parameters which significantly affect the abrasive wear behavior.

3.2.3.1.6 Methodology to study optimum controlling parameters of three-body abrasive wear

In the present experimental investigation, the controlling parameters as tabulated in Tab. 3.4 were considered as factors which influence the three-body abrasive wear of hybrid friction composites. Three-body abrasive wear test was conducted using RWAT rig according to control factors and their levels using the L_{18} orthogonal array are listed in Tab. 3.5. The specific wear rate (K_S) was calculated from equation (3.1).

Tab. 3.4: Control factors and their levels in three-body abrasive wear.

Factor	Description	Level I	Level II	Level III
A	Abrasive sand size (µm)	212	425	–
L	Applied load (N)	30	40	50
F	Fiber content (wt%)	15	20	25
D	Abrading distance (m)	1000	1500	2000

Tab. 3.5: Experimental design of L_{18} orthogonal array.

Exp. No.	Experimental parameters				Number of experiments
	Abrasive size (μm)	Applied load (N)	Fiber content (wt%)	Abrading distance (m)	
1	212	30	15	1000	18
2	212	30	20	1500	
3	212	30	25	2000	
4	212	40	15	1000	
5	212	40	20	1500	
6	212	40	25	2000	
7	212	50	15	1500	
8	212	50	20	2000	
9	212	50	25	1000	
10	425	30	15	2000	
11	425	30	20	1000	
12	425	30	25	1500	
13	425	40	15	1500	
14	425	40	20	2000	
15	425	40	25	1000	
16	425	50	15	2000	
17	425	50	20	1000	
18	425	50	25	1500	

3.2.3.1.7 Fade and recovery test

The fade and recovery behavior of the friction composites were assessed on a chase friction test rig according to IS 2742 Part 4 (SAE J661) standard. The sample size of 25 mm × 25 mm (625 mm^2, Fig. 3.5) was cut approximately equidistant from each end from the center of the cured composite and used for this experiment [23]. The detailed specifications of the counterpart drum used in the chase friction test are tabulated in Tab. 3.6 and the experimental procedure is listed in Tab. 3.7.

Fig. 3.5: Sample used for chase friction test.

Tab. 3.6: Specifications of the cast iron drum as per IS 2742 used in chase friction test.

S. No.	Characteristics	Grade FG300	
1	Composition (wt%)	Carbon	3.30–3.50
		Silicon	1.80–2.10
		Manganese	0.55–0.75
		Phosphorus	0.20 max.
		Sulphur	0.20 max.
		Nickel	0.60–0.70
		Chromium	0.15–0.25
		Molybdenum	0.20–0.30
2	Drum diameter (mm)	279	
3	Drum wall thickness (mm)	11	
4	Hardness and metallography	179–229 HBS and Pearlite grain structure	

Tab. 3.7: Experimental procedure of fade and recovery cycle.

Stages	Speed	Temperature (°C)		Load	On time		Off time	No. of Applications	Heater
	(rpm)	Min	Max	(N)	(min)	(s)	(s)		
Burnish	308	–	93	440	20			1	Off
Baseline-I	411	82	104	660		10		20	Off
Fade-I	411	82	289	660	10			1	On
Recovery-I	411	261	93	660		10		1	Off
Wear	411	193	204	660		20	10	100	Off
Fade-II	411	82	345	660	10			1	On
Recovery-II	411	317	93	660		10		1	Off
Baseline-II	411	82	104	660		10	20	20	Off

3.2.3.2 Scanning electron microscopy

The worn surface of the composite samples was examined using a scanning electron microscope (SEM). The samples were coated by a thin gold film sputter prior to examination.

3.3 Results and discussion

3.3.1 Thermal, mechanical and thermomechanical analysis of basalt-recycled aramid fiber reinforced friction composites

3.3.1.1 Thermal and mechanical properties of hybrid friction composites

The thermal and mechanical properties of fiber reinforced hybrid friction composites are listed in Tab. 3.8 The density of the friction composites decreased linearly with an

Tab. 3.8: Physical and thermal properties of hybrid friction composites.

S. No.	Composite/Properties	Unit	C1	C2	C3	C4	C5
1	Density	g/cm^3	1.76	1.82	1.82	1.83	1.84
2	Water swelling	mm	0.005	0.003	0.002	0.001	nil
3	Heat swelling	mm	0.16	0.14	0.13	0.10	0.07
4	Acetone extraction	%	1.16	1.21	1.24	1.57	1.93
5	Loss of weight on ignition at 800 °C	%	12	16	19	20	22
6	Hardness	(Shore-D)	85.2	84.4	80.5	79.6	72.2
7	Tensile strength, σ	MPa	12.85	11.47	8.13	9.23	5.96
8	Elongation, e	mm	1.15	1.1	1.06	0.86	0.56
9	Compressive strength	MPa	66.96	54.32	43.89	39.77	27.92
10	Flexural strength	MPa	50.09	47.99	41.33	34.56	24.22
11	Impact energy	Joules	0.451	0.263	0.359	0.292	0.151
12	Impact strength	kJ/mm^2	7.54	5.69	6.00	4.77	3.23

increase in wt% of fiber content, which is due to basalt fiber having a lower level of density (2.65 g/cm^3) when compared to the density of $BaSO_4$ (4.5 g/cm^3). The water and heat swelling of the friction composites increase with an increase in fiber content from 5 to 25 wt%, which shows that fiber content has a higher water and heat absorption capacity. The acetone extraction and loss of weight on ignition of the composite samples decrease with an increase in wt% of fiber content, which indicates that the developed composite C1 has less uncured resin since it has more porosity and high fiber content, which in turn gives way and allows the curing heat to penetrate through the materials, thus curing the composites effectively. Loss of weight on ignition of the friction composites at 800 °C decreases from 12 to 22 wt% with increasing the fiber content of the composites from 5 to 25 wt%. This indicates that composites with lower fiber content contribute to increasing the thermal degradation due to heat, indicating their poor thermal stability.

The composite C1 with higher amount of fiber (25 wt%) is 1.2 times harder than composite with a lower fiber content (C5, 5 wt%). This is because the hard basalt fibers in the composites are basically composed of minerals, plagioclase, pyroxene, and olivine and have comet-like microstructures with a tail and a head or 'shot' morphology [11]. Since the 'shot' in the basalt fiber is hard it exhibits a higher level of resistance to indentation and restricts the plastic deformation of the hybrid friction composites. The relative uniform distribution of all phases and decreased interparticle distance within the matrix result in increased resistance of composites to indentation. The composite with an increased amount of fiber content exhibits the highest tensile strength (12.85 MPa) when subject to a tensile load. The result proves that increasing the wt% of fiber from 5 to 25 wt% results in a considerable increase in tensile strength of friction composites. Also, micropacking of fibers in the cured phenol formaldehyde matrix forms a crosslinked network structure which causes mechanical compaction of the fibrous ingredients and also shows effective synergism.

From Fig. 3.6 (a)–(c), the good adhesion between the fiber-matrix interfaces can be confirmed. The fiber-matrix interfaces show better surface integrity in composite C1 and a minimum amount of fiber pull-out and fiber bonding (Marked as B) in the matrix can be noticed. This is because composite C1 possesses high interfacial stiffness corresponding to a high composite strength/modulus. The fiber pull-out, fiber thinning and generation of more voids decrease the tensile strength of lower wt% of composite (C5, 5 wt%), which was confirmed by the SEM studies. The higher fiber content in the friction composites eliminates the fiber crushing under compressive stress and imparts better bending resistance and impact strength to the hybrid friction composites. In the case of commercial friction material, the maximum compressive strength was 110 MPa [24]. The superior mechanical properties of basalt fiber and the fibrillation, ductility and elongation properties of recycled aramid fiber increase the stress transfer between the fiber and matrix and strengthen the fiber-matrix interface to enhance the flexural strength of composite C1. Composites with a higher amount of fiber content show better impact energy, which varies from 0.151 to 0.451 Joules with an increase in fiber content from 5 wt% (C5) to 25 wt% (C1). The gradual increase in impact energy is due to the fracture strength of fibers (basalt+recycled aramid) and compression pressure which eliminates void content in the friction composites.

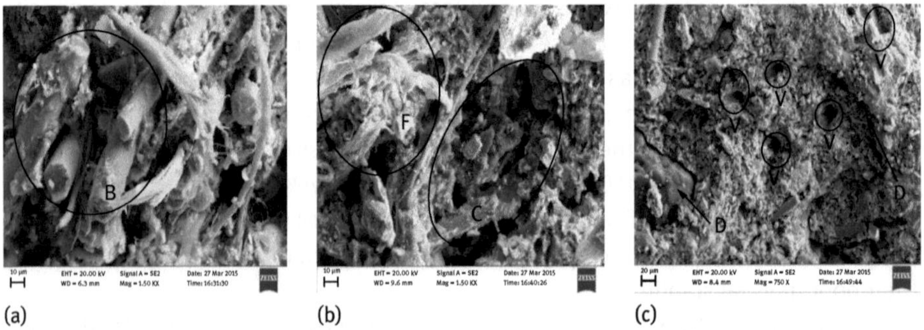

Fig. 3.6: SEM photomicrograph of fractured surface of composite samples (a) C1, (b) C3, and (c) C5.

3.3.1.2 Thermogravimetric analysis

Thermogravimetric analysis (TG/DTG) is another important thermomechanical analysis that is used to study the thermal degradation of carbonaceous materials and measures the volatile components present in the friction composites. It is used to measure the rate of weight loss of composites as a function of temperature. The weight changes could be due to desorption of volatile components from the composite samples or absorption of gaseous components from the atmosphere surrounding the composite samples when heat is supplied. Fig. 3.7 shows the thermal stability observation of basalt fiber and Fig. 3.8 (a)–(e) demonstrates the TG/DTG curves of fabricated friction composite samples. The TG/DTG results for the hybrid friction composites heated at a rate of 20 °C/min under air atmosphere from 30 to 800 °C are shown in Tab. 3.9. The percentage of weight loss of C1 was 33.32 and 38.57, 40.43, 40.82, and 47.23 for composites C2, C3, C4, and C5 respectively. This proves that the thermal stability is higher and optimal for composite C1 when compared to all other formulations and incorporation of basalt fiber enhances the thermal stability of friction composites in a beneficial way.

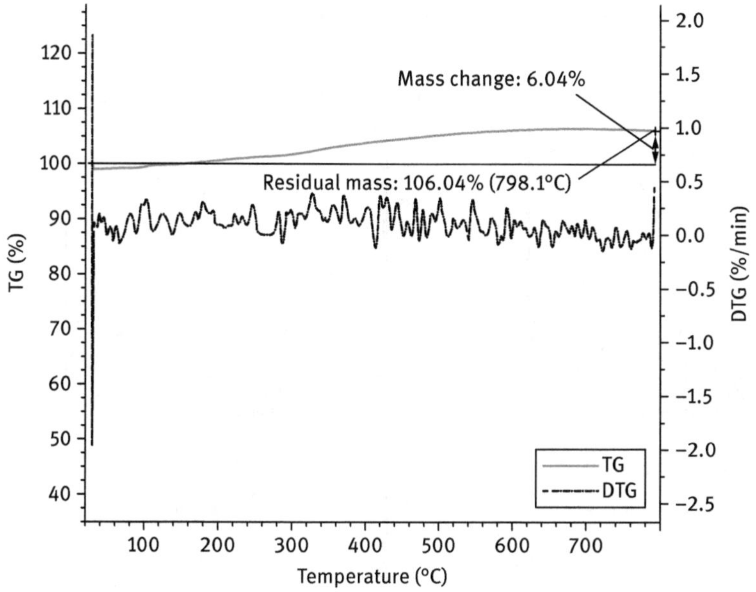

Fig. 3.7: TG/DTG observations of basalt fiber.

Tab. 3.9: TG/DTG results of hybrid friction composite.

Composites	I Wt loss (%)		II Wt loss (%)		III Wt loss (%)	
	Temp. range (°C)	Wt loss (%)	Temp. range	Wt loss (%)	Temp. range	Wt loss (%)
C1	30–295	0.64	295–398	7.69	398–485	14.4
C2	30–142	2.16	142–360	4.61	360–472	19.11
C3	30–261	1.84	261–355	6.9	355–488	23.52
C4	30–248	2.54	248–345	8.4	345–472	21.78
C5	30–257	1.87	257–379	9.19	379–491	22.82

Tab. 3.9: (continued)

Com-posites	IV Wt loss (%)		Total Wt loss	Residual mass
	Temp. range (°C)	Wt loss (%)		
C1	485–798	10.59	33.32	66.68
C2	472–798	12.69	38.57	61.43
C3	488–798	8.08	40.43	59.57
C4	472–798	8.1	40.82	59.18
C5	491–798	13.35	47.23	52.77

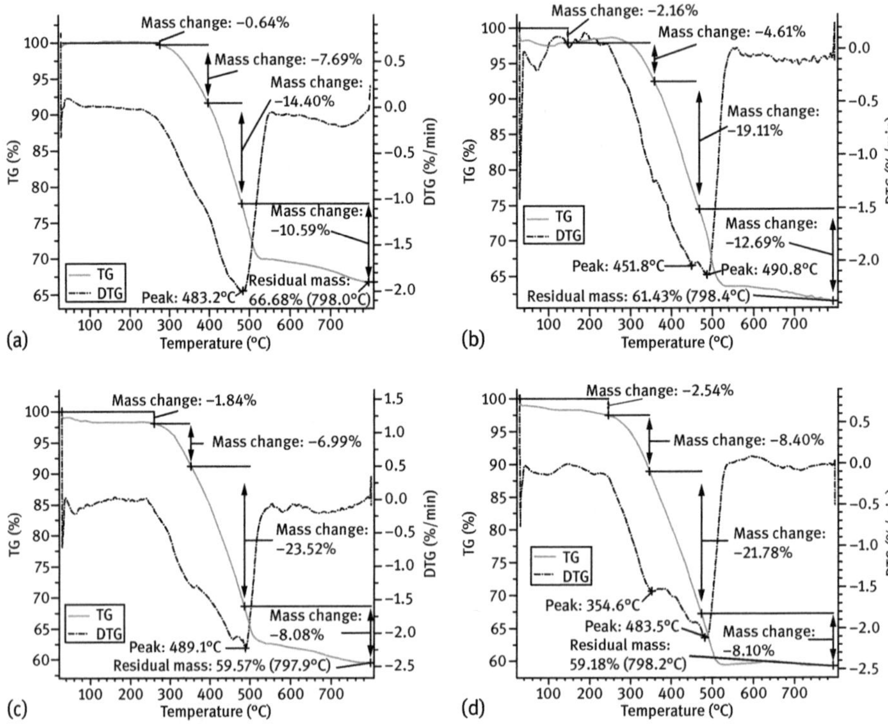

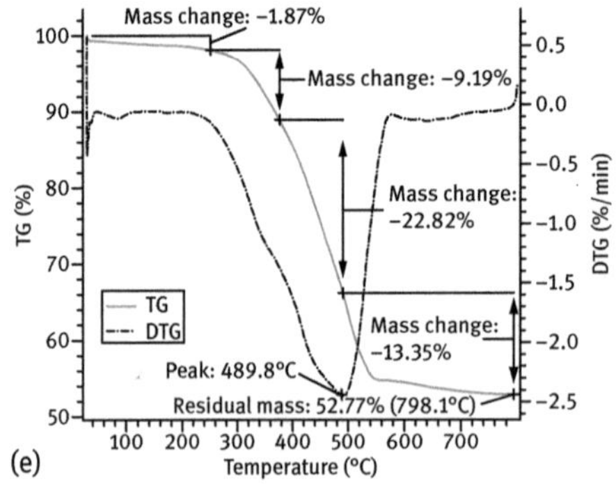

Fig. 3.8: TG/DTG observations hybrid friction composite (a) C1, (b) C2, (c) C3, (d) C4, and (e) C5.

3.3.2 Abrasive wear behavior of basalt-recycled aramid fiber reinforced hybrid friction composites

During braking, abrasion is caused at the friction interface, either by hard wear debris particles present in the brake friction lining or by entrapment of sand particles between friction composite and counterface disc. Hence, it is necessary to understand the influence of reinforcements and abrasive particles on abrasive wear behavior of friction composites. Also, the optimization of abrasive wear parameters is essential to observe the relation between the properties of abrasive particles and the wear they cause.

3.3.2.1 Influence of fiber reinforcement and abrasive particle size on three-body abrasive wear of hybrid friction composites

Three-body abrasive wear behavior of hybrid friction composites was tested using a rubber wheel abrasion test (RWAT) rig. The abrasive wear tests were conducted for the friction composite C1, C2, C3, C4, and C5 specimens; the wear scratch on samples after the experiments was indicated as shown in Fig. 3.9. The wear experiments were conducted for the composite samples as a function of increasing abrading distance (250 to 1000 m) under a constant applied load of 40 N for different abrasive particle sizes (212 and 425 μm) at 200 rpm.

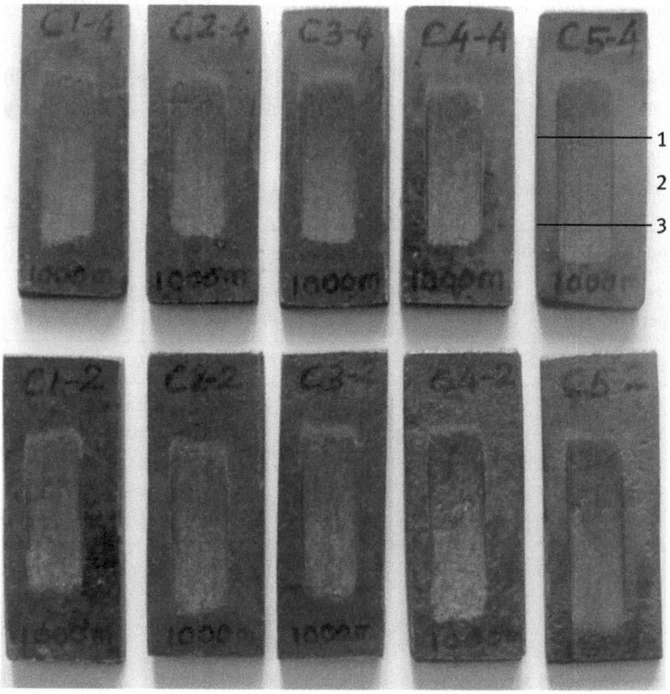

Fig. 3.9: A typical wear scratch of hybrid friction composites.

3.3.2.1.1 Abrasive wear volume loss

The variations in abrasive wear volume loss of hybrid friction composites are shown in Tab. 3.10. The wear volume loss data is shown in Fig. 3.10 (a) and (b). The wear loss of the composites tends to increase almost linearly with increasing abrading distance and abrasive particle size. The wear volume loss due to abrasion was maximum for composite C5 with a minimum amount (5 wt%) of reinforcement fiber (basalt). It gradually decreased with increasing fiber content and it was minimum for composite C1 with a maximum amount (25 wt%) of reinforcement fiber (basalt + recycled aramid). The results are in good agreement with those observed by Chand et al. [8] that the loss of wear volume increases with an increasing abrading distance and is greater for coarse abrasives when compared to fine silica sand abrasives. This is because the coarse abrasive particles are relatively hard and can penetrate deeply into the matrix compared to fine abrasives that remove the material from the surface at a faster rate. When the friction composites are abraded under the coarse abrasive particles, the number of abrasive particles in contact with the composite surface area is smaller. Hence, higher stresses are developed for the same applied load resulting in a higher removal rate of material. However, for finer abrasives, the area of contact with the friction composite surface is larger and elastic in nature, thus supporting the same load without the removal of more material. Also, the fine abrasives situate more closely and the applied load is shared by a large number of particles in contact. Hence, the energy gain or load carried per fine particle is minimal, resulting in insufficient energy to facilitate failure in the composite. In three-body abrasion, the abrasive particles are free to roll or slide and these free-falling abrasives gain energy from the high speed rubber wheel after contact with it and strike the composite, resulting in the formation of pits or plastic deformation of the matrix. In addition, abrasive particles roll between the composite and the rubber wheel, causing plastic deformation to the composite.

Tab. 3.10: Wear volume loss of hybrid friction composites.

Sample designation	Wear volume loss × 10^3 mm^3							
	Abrasive size (212 μm) Abrading distance (m)				Abrasive size (425 μm) Abrading distance (m)			
	250	500	750	1000	250	500	750	1000
C1	0.0937	0.1252	0.1580	0.1780	0.1329	0.1750	0.2354	0.3026
C2	0.1206	0.1727	0.2101	0.2555	0.1643	0.2265	0.3047	0.3750
C3	0.1620	0.2109	0.2555	0.2992	0.2007	0.2853	0.3738	0.4707
C4	0.2176	0.3084	0.4137	0.5011	0.2538	0.3731	0.4897	0.6319
C5	0.2739	0.4457	0.5795	0.7085	0.3589	0.5152	0.6653	0.8212

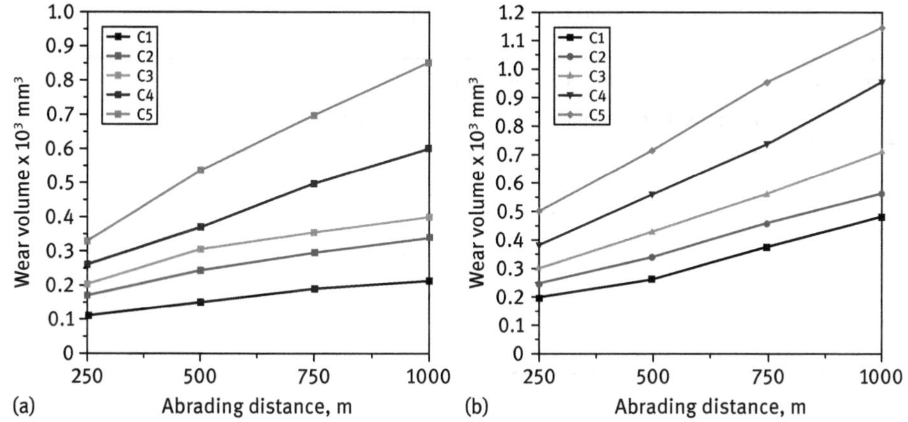

Fig. 3.10: Effect of abrading distance on wear volume loss of friction composites against: (a) 212 μm and (b) 425 μm silica sand at 40 N.

3.3.2.1.2 Specific wear rate

The specific wear rates as a function of increasing abrading distance for two different abrasive particle sizes (212 and 425 μm) at a constant load of 40 N are listed in Tab. 3.11. Fig. 3.11 (a) and (b) show histograms of comparative specific wear rates (K_S) of friction composites under varying abrading conditions. Initially, the specific wear rate is high for all composite samples, but gradually decreases with increasing abrading distance from 250 to 1000 m and then tends to become steady after a certain abrading distance. This is due to the fact that the removal of polymer matrix exposes the reinforcement fibers and hard fillers that further resist material removal by acting as a protective layer. Similar examinations were made by Basawarajappa et al. [25] in the case of G-E composite filled with SiC particles and they reported that when the abrading distance increases, the hard SiC particulates are exposed and resist the penetration of sand particles into the composites. Hence the wear rate of the composites is reduced over increasing abrading distance. Moreover, the linear wear hypothesis revealed a close approximation of wear performance of friction material with predicted pad life [4].

In general, the mechanical and tribological properties of the hybrid friction composites are mostly governed by the type and amount of reinforcement fibers and fillers. The hardness of the composite is one of the factors which determine the abrasive wear process [26]. Basalt fiber exhibits a high specific modulus, hardness, excellent fiber/matrix adhesion and abrasion resistance that could be related to the degree of crystallinity and orientation of fibers [13]. Also, basalt fiber has a tail-and-head or shot morphology, the "shot" being the harder part of the basalt fiber [11]. In the case of hybrid friction composites, the addition of basalt fiber along with ceramic fiber decreases the specific wear rate of the composites [10]. Superior wear resistance, good filler retention and nonabrasiveness to mating surfaces are some of the key contributions of

Tab. 3.11: Specific wear rate of hybrid friction composites.

Composite sample designation	Specific wear rate × 10^{-11} m^3/Nm							
	Abrasive size (212 μm) Abrading distance				Abrasive size (425 μm) Abrading distance			
	250	500	750	1000	250	500	750	1000
C1	0.9370	0.6260	0.5267	0.4450	1.3288	0.8753	0.7848	0.7564
C2	1.2062	0.8636	0.7004	0.6389	1.6432	1.1324	1.0157	0.9374
C3	1.6198	1.0544	0.8516	0.7479	2.0072	1.4264	1.2496	1.1768
C4	2.1760	1.5410	1.3790	1.2528	2.5376	1.8656	1.6323	1.5798
C5	2.7390	2.2285	1.9317	1.7713	3.5890	2.5760	2.2217	2.0530

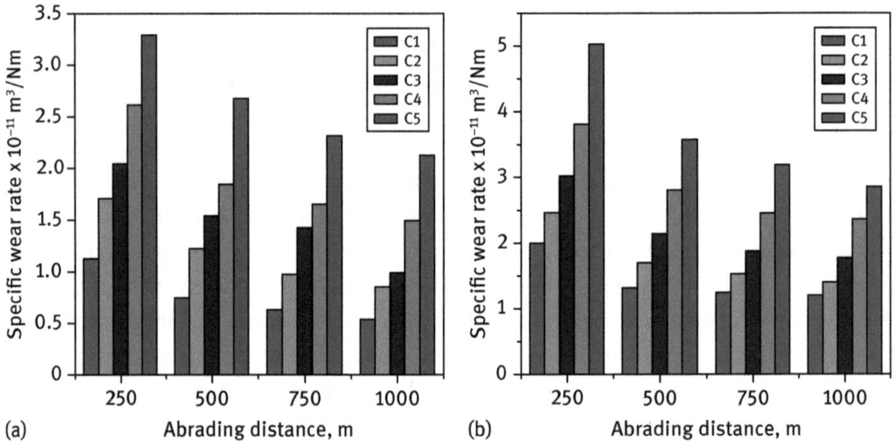

Fig. 3.11: Effect of abrading distance on specific wear rate of friction composites against: (a) 212 μm and (b) 425 μm silica sand abrasive at 40 N.

aramid fiber, which makes it an essential reinforcement in friction composites when the composites are subjected to abrasive wear conditions. During abrasion, the aramid fiber improves the wear resistance of the composite due to its characteristic properties of fibrillation and elongation [16]. Similarly, Kato and Magario [15] evaluated the role of aramid fibers in the wear of composite brake pads and confirmed that the addition of aramid fiber considerably improves the wear resistance of the composite pad. As the load is applied, the stress concentration is accumulated around the hard fibers and fillers of the friction composites. Also, the depth of penetration of the abrasives depends on the type and interparticle distance between the reinforcements in the friction composites. Hence, the higher amount of fiber reinforcements in the composite C1 reduces the interparticle distances and provides more obstruction to abrasives trying to penetrate into the composite.

The phenomenon of decreasing specific wear rate as a function of increasing distance is due to the basic structure and the hardness of hard reinforcement fibers and fillers dispersed into the soft phenol formaldehyde matrix. This indicates that at the initial stage of abrasion, the abrasive particles are in contact with phenol formaldehyde matrix that has lesser hardness when compared to angular silica sand. At this instance, the ratio of H_a (hardness of abrasives) to H_s (hardness of surface) is much more than unity, which results in severe matrix damage and a higher material removal rate. Similarly, when abrading distance is increased, an increased number of reinforcement fibers along with hard dispersoid particles are exposed and come into contact with the abrasive sand particles, which results in better abrasion resistance of friction composites. Hence, the H_a/H_s is less than unity and thus high energy is required or abrasive particles have to work harder to facilitate failure in fibers [8]. Therefore, the rate at which material removal occurs with respect to abrading distance decreases. Similar observations were made by Suresha et al. [27] in the case of three-body abrasive wear behavior of C-E and G-E composites and they concluded that the loss of wear volume increased linearly with increasing abrading distance, but the specific wear rate decreased with an increase in abrading distance.

With an increase in size of the abrasive particles from 212 to 425 µm, the specific wear rate is increased. This is because the coarse abrasive particles are harder and retain their shape and angularity for a longer duration of time, thus effectively removing the material at a faster rate by extensive microcutting, microcracking, and microfatigue and microplowing mechanisms. This results in the formation of debris or microchips and also causes localized brittle fracture of the composite material [9]. These microcracks freely propagate around the wear groove, resulting in excessive material removal by fragmentation. Also, for a constant applied load, the effective stress on individual abrasives increases with coarser abrasive particles, since the load is shared by a smaller number of abrasive particles. Hence, the microcutting and microplowing actions performed due to the higher stress on the composite surface cause more material removal when the composites are abraded under coarser abrasives.

3.3.2.1.3 Morphology of worn surfaces

To substantiate the wear data, scanning electron microscopy (SEM) photomicrographs of abraded samples of C1, C3, and C5 are shown in Fig. 3.12 (a) and (b) to 3.14 (a) and (b). They were subjected to a higher abrading distance of 1000 m at a constant load of 40 N for different sizes of abrasives. The worn surface of composite C1 at 1000 m distance using angular silica sand of 212 µm is shown in Fig. 3.12 (a). The abraded surface is relatively smooth with less evidence of fiber fragmentation and matrix damage when compared to Fig. 3.12 (b). The minimal and localized generation of voids (marked as V) due to pull-out of fibers after repeated abrasion are evident in Fig. 3.12 (a). It appears to have a better adhesion between the fiber-matrix interfaces that obstruct the removal of matrix during abrasion, resulting in a worn surface that is relatively smooth with fewer microcracks and voids, shown in Fig. 3.12 (a). However, under the action of coarse silica sand as abrasive, the severity of damage to the abraded surface of the composite is relatively higher when compared to the abraded surface under fine abrasive particles, as shown in Fig. 3.12 (b). The wear resistance offered by reinforcement fibers and hard filler particles improve the surface integrity, hence the minimal extent of microcracks (marked as C), fiber debonding, high degree of fiber breakages and broken fibers (marked as B) are noticed predominantly along the direction of flow of abrasives as shown in Fig. 3.12 (b). Also, surface fatigue under repeated abrasion is another reason for fiber pull-out and void generation in the composites. As a result, the surface is left with imprints (marked as I) of separating fibers, displaced fibers (marked as B) and the appearance of shallow furrows due to the microplowing mechanism.

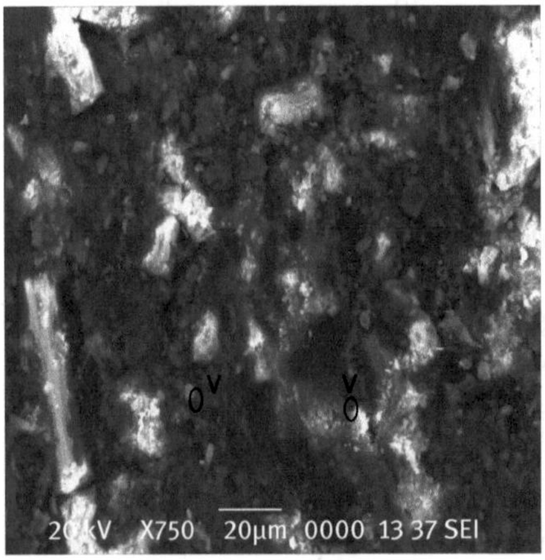

(a)

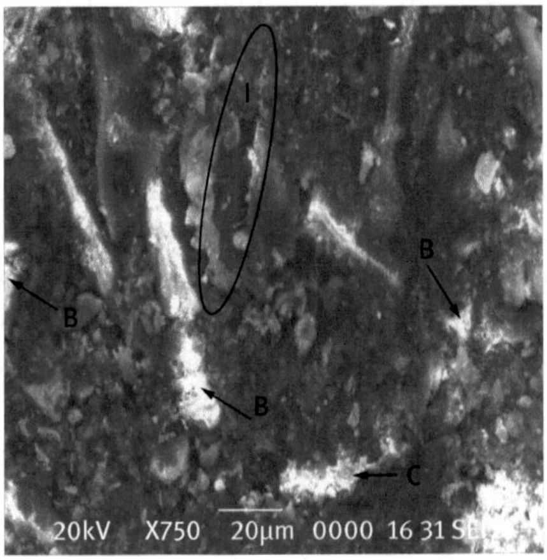

(b)

Fig. 3.12: Worn surface photographs of friction composite C1 at 1000 m distance against: (a) 212 μm and (b) 425 μm silica sand abrasives at 40 N.

Fig. 3.13 (a) describes the effect of fine abrasive particles on composite C3. The matrix is distorted and damaged due to the entrapment of the abrasive particles between the rubber wheel and the specimen, which induces a high stress in the particles resulting in microplowing with a large amount of wear debris (marked as D) formation that can be noticed in Fig. 3.13 (a). Thinning of the matrix is caused by fracture of the fibers which are forced out of the matrix, forming deep pits (marked as P), as evident in Fig. 3.13 (a). Deep plowed furrows (marked as F), which are generated by brittle fracture, are evident in Fig. 3.13 (b). Fragments of fiber torn (marked as T) or peeling off from the matrix is another remarkable feature that occurs due to the abrasion and transverse cutting (marked as G) effects of sharp abrasive particles. Interfacial debonding caused poor adhesion between reinforcements and phenol formaldehyde matrix resulting in accelerated breakage of matrix interspersed with plowed furrows corresponding to the direction of wear, which are evident in Fig. 3.13 (b). The ductility and elongation of aramid fibers were found to improve the wear behavior of composites C1 and C3 substantially, which is not apparent in C5, resulting in severe wear among the formulated composites as shown in Fig. 3.14 (a) and (b). The low modulus phenol formaldehyde matrix is severely damaged and removed quickly due to plowing action of abrasives with resultant deep furrows (marked as F), as shown in Fig. 3.14 (a). In Fig. 3.14 (b), long microcracks (marked as L) are evident in the longitudinal direction

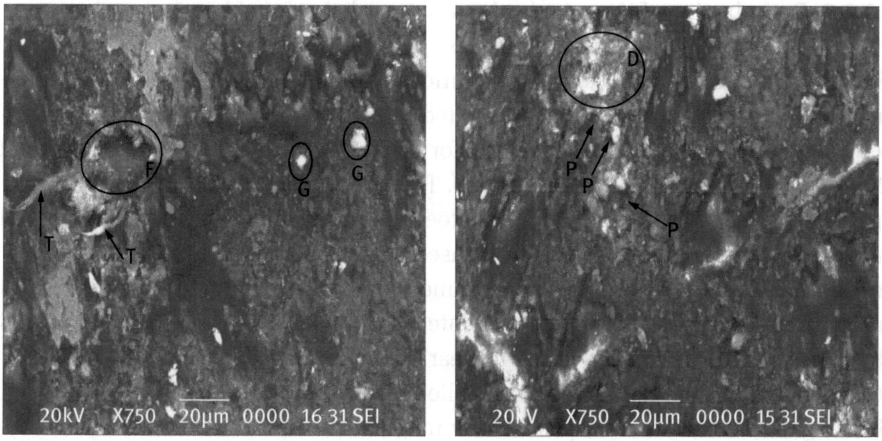

(a) (b)

Fig. 3.13: Worn surface photographs of friction composite C3 at 1000 m distance against: (a) 212 and (b) 425 μm silica sand abrasives at 40 N.

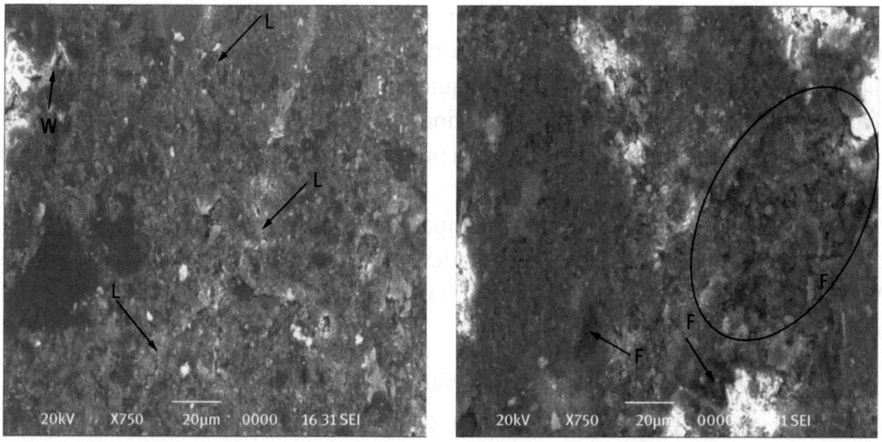

(a) (b)

Fig. 3.14: Worn surface photographs of friction composite C5 at 1000 m distance against: (a) 212 and (b) 425 μm silica sand abrasives at 40 N.

of the worn surface due to surface fatigue by repeated abrasion. A governing feature observed on the surface of C5 was the generation of wide cracks (marked as W) in the abrading direction as shown in Fig. 3.14 (b). When the network of cracks intersects, the filler particle tends to become loose and is removed in the form of wear debris. Thus a higher degree of wear in composite C5 can be attributed to lower matrix ductility and poor fiber reinforcement, which deform the matrix by surface disintegration.

3.3.2.2 Optimization of three-body abrasive wear behavior of basalt-recycled aramid fiber reinforced friction composites

Three-body abrasive wear is affected by various parameters. From the detailed experimentation on the friction composites tested under three-body abrasive wear conditions, better performance was obtained for the composites C1 (25 wt% of fiber), C2 (20 wt% of fiber), and C3 (15 wt% of fiber). This experimental work aims at minimizing the specific wear rate (K_S) and wear coefficient and at maximizing hardness (μ) of hybrid friction composites. The grey-based Taguchi approach is used for the composites C1, C2, and C3 in order to determine the optimal parameter combination to achieve minimization of abrasive wear related to tribological applications. The results obtained from the three-body abrasive wear and mechanical tests are used to evaluate the influence of abrasive size (A), applied load (L), wt% of fiber content (F), and abrading distance (D) on the specific wear rate, wear coefficient, and hardness of hybrid friction composites are tabulated as Tab. 3.12.

3.3.2.2.1 Optimization of abrasive wear test parameters using grey relational analysis

A lower value of specific wear rate and wear coefficient and a higher value for hardness are identified as the indicators of best performance under three-body abrasive wear conditions. For the data preprocessing, equation (3.3) is used to evaluate the specific wear rate and wear coefficient while equation (3.4) is used to evaluate the hardness of the friction composites. The values of data preprocessing (normalized data) are listed in Tab. 3.13.

All the sequences (reference and comparability) after data preprocessing using equation (3.5) are listed in Tab. 3.13 The deviation sequences for specific wear rate ($\Delta_{0i}(1)$), wear coefficient ($\Delta_{0i}(2)$) and for hardness ($\Delta_{0i}(3)$) can be calculated as follows:

$$\Delta_{03}(1) = |x_0^*(1) - x_3^*(1)| = |1.00 - 0.9608| = 0.0392 \text{ (for specific wear rate)},$$
$$\Delta_{03}(2) = |x_0^*(2) - x_3^*(2)| = |1.00 - 0.9577| = 0.0423 \text{ (for wear coefficient)},$$
$$\Delta_{02}(3) = |x_0^*(3) - x_3^*(3)| = |1.00 - 1.0000| = 0 \text{ (for hardness)}.$$

Hence $\Delta_{02} = (0.0392, 0.0423, \text{ and } 0)$.

Tab. 3.12: Results of three-body abrasive wear of L_{18} orthogonal array.

Exp. No.	Experimental Parameters				Optimizing Parameters		
	Abrasive size (μm)	Applied load (N)	Fiber content (wt%)	Abrading distance (m)	Specific wear rate (mm^3/Nm)	Wear co-efficient	Hardness (Shore-D)
1	212	30	15	1000	0.0059	0.4791	81
2	212	30	20	1500	0.0051	0.4334	84
3	212	30	25	2000	0.0036	0.3094	85
4	212	40	15	1000	0.0075	0.6058	81
5	212	40	20	1500	0.0053	0.4499	84
6	212	40	25	2000	0.0040	0.3370	85
7	212	50	15	1500	0.0055	0.4427	81
8	212	50	20	2000	0.0033	0.2798	84
9	212	50	25	1000	0.0052	0.4450	85
10	425	30	15	2000	0.0060	0.4896	81
11	425	30	20	1000	0.0069	0.5815	84
12	425	30	25	1500	0.0049	0.4144	85
13	425	40	15	1500	0.0110	0.8946	81
14	425	40	20	2000	0.0070	0.5870	84
15	425	40	25	1000	0.0077	0.6515	85
16	425	50	15	2000	0.0099	0.8003	81
17	425	50	20	1000	0.0116	0.9795	84
18	425	50	25	1500	0.0085	0.7225	85

Tab. 3.13: Data preprocessing and deviation sequence of hybrid friction composites.

Exp. No.	Data preprocessing			Deviation sequence		
	Specific wear rate	Wear coefficient	Hardness	Specific wear rate $\Delta_{0i}(1)$	Wear coefficient $\Delta_{0i}(2)$	Hardness $\Delta_{0i}(3)$
1	0.6864	0.7151	0.0000	0.3136	0.2849	1.0000
2	0.7805	0.7805	0.8500	0.2195	0.2195	0.1500
3	0.9608	0.9577	1.0000	0.0392	0.0423	0.0000
4	0.4977	0.5341	0.0000	0.5023	0.4659	1.0000
5	0.7569	0.7569	0.8500	0.2431	0.2431	0.1500
6	0.9216	0.9182	1.0000	0.0784	0.0818	0.0000
7	0.7407	0.7672	0.0000	0.2593	0.2328	1.0000
8	1.0000	1.0000	0.8500	0.0000	0.0000	0.1500
9	0.7684	0.7639	1.0000	0.2316	0.2361	0.0000
10	0.6707	0.7001	0.0000	0.3293	0.2999	1.0000
11	0.5688	0.5688	0.8500	0.4312	0.4312	0.1500
12	0.8118	0.8076	1.0000	0.1882	0.1924	0.0000
13	0.0676	0.1212	0.0000	0.9324	0.8788	1.0000
14	0.5609	0.5609	0.8500	0.4391	0.4391	0.1500
15	0.4753	0.4687	1.0000	0.5247	0.5313	0.0000
16	0.2081	0.2561	0.0000	0.7919	0.7439	1.0000
17	0.0000	0.0000	0.8500	1.0000	1.0000	0.1500
18	0.3745	0.3673	1.0000	0.6255	0.6237	0.0000

The same calculation method was carried out for $i = 1, 2, \ldots, 18$ and deviation sequences calculated are shown in Tab. 3.13. Tab. 3.14 shows the calculated values of grey relational coefficient (GRC) for each experiment of the L_{18} orthogonal array by applying equation (3.6) for individual quality characteristics of specific wear rate, wear coefficient and hardness of the friction composites. After calculating GRC, the grey relational grade has been computed by averaging the grey relational coefficient corresponding to each performance characteristic by using equation (3.7). Grey relational grade (GRG) represents the level of correlation between the reference sequence and the comparability sequence. The numerical values of GRG between elements are not significant, while the grey relational orders between the elements provide more information.

Tab. 3.14: GRC and GRG of hybrid friction composites.

Exp. No.	GRC			GRG	Rank
	Specific wear rate	Wear coefficient	Hardness		
1	0.6145	0.6370	0.3333	0.5283	13
2	0.6949	0.6949	0.7692	0.7197	6
3	0.9273	0.9220	1.0000	0.9497	1
4	0.4989	0.5176	0.3333	0.4499	16
5	0.6729	0.6729	0.7692	0.7050	7
6	0.8644	0.8594	1.0000	0.9079	3
7	0.6585	0.6823	0.3333	0.5580	12
8	1.0000	1.0000	0.7692	0.9231	2
9	0.6834	0.6793	1.0000	0.7876	5
10	0.6029	0.6250	0.3333	0.5204	14
11	0.5369	0.5369	0.7692	0.6144	10
12	0.7266	0.7222	1.0000	0.8162	4
13	0.3491	0.3626	0.3333	0.3483	18
14	0.5324	0.5324	0.7692	0.6114	11
15	0.4879	0.4848	1.0000	0.6576	8
16	0.3870	0.4020	0.3333	0.3741	17
17	0.3333	0.3333	0.7692	0.4786	15
18	0.4443	0.4414	1.0000	0.6286	9

A higher value of GRG is considered as the strongest relational degree between both the reference and the comparability sequence and this GRG value corresponds to an optimal parameter combination. It is clear that a lesser abrasive wear behavior of a friction composite is offered by a comparability sequence with a higher value of GRG. From the combined factor results, the highest GRG is assigned the highest order, while the minimum GRG is assigned the lowest order. The ordering (ranking) of the GRG values is shown in Tab. 3.14. The optimal parameter setting for the multiple performance of minimum specific wear rate, wear coefficient and maximum hardness is obtained for experiment number 3, which has the highest grey relational grade (0.9497). This implies that the factor setting of experiment number 3 provides the nearest optimal setting from the point of view of three-body abrasive wear. Since the factorial design is orthogonal, the effect of each parameter on the GRG value at various levels can be separated. For example, the mean of GRG for factor *A* at two levels can be evaluated by calculating the average of the GRG for experiments 1 to 9 and 10 to 18. Similarly, the mean of the GRG for each level of other parameters was calculated and is listed in Tab. 3.15. The total mean of GRG of this orthogonal array can be obtained by dividing the sum of GRG with the total number of experiments, as summarized in Tab. 3.15. The delta (max–min) value has been computed as the difference between the highest and lowest value among the values in each row and based on this, the Tab. 3.15 assigns rank for optimum parameter setting.

Tab. 3.15: Response table of grey relational grade.

Symbol	Parameter	Grey Relational Grade			Delta (max–min)	Rank
		Level 1	Level 2	Level 3		
A	Abrasive size (μm)	0.7255	0.5611		0.1644	2
L	Applied load (N)	0.6915	0.6134	0.6250	0.0781	4
F	Fiber content (wt%)	0.4632	0.6754	0.7913	0.3281	1
D	Abrading distance (m)	0.5861	0.6293	0.7144	0.1284	3
	Total mean of grey relational grade = 0.6433					

In general, the large difference in the GRG from one factor setting to another implies that the specific design parameter is an important factor in reaching the performance characteristic. It is clear from Tab. 3.15 that the delta values of GRG of abrasive wear parameters are as follows: 0.1644 for abrasive particle size, 0.0781 for applied load, 0.3281 for fiber content and 0.1284 for abrading distance. The maximum value for fiber content viz. 0.3281 indicates that of the four parameters, the fiber content (factor F) has the strongest effect on the multiresponse characteristics in controlling the abrasive wear behavior of hybrid friction composites under investigation. This shows that abrasive wear behavior is strongly determined by the addition of fiber content (*F*), followed by abrasive particle size (*A*), abrading distance (*D*) and applied load (*L*) in that order of priority. A similar trend was observed in a study by Ramesh and Suresha [28] on abrasive wear mode of fabric-reinforced C-E composites filled with functional fillers. They concluded that filler content is the most influential factor in determining the abrasive wear behavior of composites followed by abrasive grit size, abrading distance and applied load. The optimum parameter combination for specific wear rate, wear coefficient and hardness can be evaluated from the values of GRG. In the present experimental study, it is $A_1L_1F_3D_3$, i.e., silica sand abrasive size of 212 μm (A_1), applied load of 30 N (L_1), composite with 25 wt% fiber content (F_3) and abrading distance of 2000 m (D_3). If the GRG is higher, the product quality will be closest to the ideal value that is desired for optimum wear performance. Fig. 3.15 illustrates the signal-to-noise (S/N) ratio of GRG for each parameter and confirms the order of priority of controlling parameters which determines the abrasive wear behaviors. Hence the findings of the present study are similar to the literature discussed above.

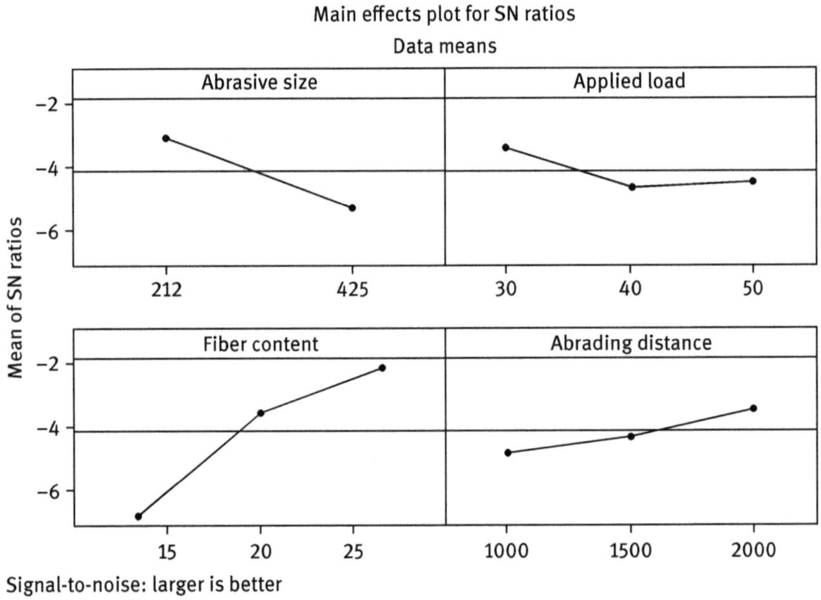

Fig. 3.15: Effect of abrasive wear parameters levels in terms of grey relational grade.

Basically, the wear is associated with either the material's properties or some other external factors. The abrasive wear behavior of the polymer based friction composite is mainly governed by the properties of the materials in contact with each other and the operating parameters. Since abrasive wear involves the cutting away of soft matrix material, viz. phenol formaldehyde matrix by a hard abrasive, it is imperative to adding hard inorganic fillers to a soft matrix to enhance the overall abrasion resistance of the composites. In an experimental work, Satapathy and Bijwe [29] studied the effect of hard alumina particles in improving the abrasive wear resistance of the phenol formaldehyde matrix-based composites.

3.3.2.2.2 Results of analysis of variance

Tab. 3.16 shows the ANOVA values for GRG which was carried out using Minitab R15. The percentage contribution of fiber content (58.85 %) was found to be the most effective parameter affecting the wear behavior of composites. The abrasive particle

Tab. 3.16: ANOVA results for combined response of specific wear rate, wear coefficient and hardness of friction composites.

Source of variation	DOF	Seq. SS	Adj. SS	Adj. MS	*F* ratio	*P* value	% of contribution
Abrasive Size	1	0.1216	0.12163	0.121634	31.9100	0	21.55
Applied Load	2	0.021302	0.0213	0.010651	2.7900	0.109	3.77
Fiber Content	2	0.332174	0.3322	0.166087	43.5800	0	58.85
Abrading Distance	2	0.051201	0.0512	0.025601	6.7200	0.014	9.07
Error	10	0.03811	0.038113	0.0038	–	–	6.75
Total	17	0.56442	–	–	–	–	100

DOF: degrees of freedom, SS: sum of squares, MS: mean square

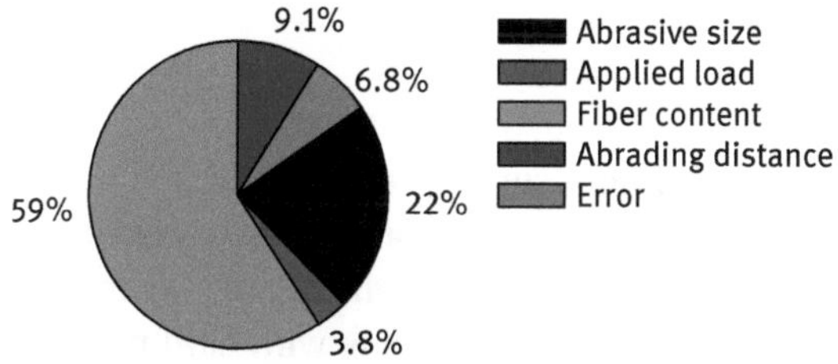

Fig. 3.16: Contribution (%) of factors on grey relational grade.

size (21.55 %) was found to be next largest influencing factor, followed by abrading distance (9.07 %) and applied load (3.77 %), as shown Fig. 3.16, which illustrates the various levels of factors considered.

3.3.2.2.3 Confirmation experiment

Generally, a verification test needs to be carried out in order to check the accuracy of analysis after optimal levels of test parameters have been identified. Once the optimum level has been selected, the next step is to predict and verify the improvement to the values for the multiple responses of the control factors affecting the wear in three-body abrasive wear condition as listed in Tab. 3.17. The predicted optimum value can be calculated by means of additive law as given by:

$$\hat{y} = y_m + \sum_{i=1}^{q} (\hat{y}_i - y_m). \tag{3.8}$$

y_m: total mean of GRG,

$\hat{y}_i$: mean of GRG at optimum level,

q: number of main control parameters that significantly affect the multiple response characteristics such as specific wear rate, wear coefficient and hardness.

Tab. 3.17: Comparison of predicted and experimental values of GRG.

	Results for confirmation of experiment	
	Experimental value	**Predicted value**
Combination of testing parameters	$A_1L_1F_3D_3$	$A_1L_1F_3D_3$
GRG	0.9497	0.9928

ANOVA clearly indicates that the abrasive particles size (A), wt% of fiber content (F), and abrading distance (D) are the significant factors and the applied load (L) is the least significant one. The predicted value of GRG is 0.9928 and the experimental value of GRG using optimized test parameters is 0.9497, which is in good agreement with the experimental value as listed in Tab. 3.17. Thus, the statistics for GRG presented in this work are in line with the experimental investigation of three-body abrasive wear of hybrid friction material composites.

3.3.2.2.4 Correlation of optimum results with worn surface morphology

The SEM photomicrographs for the worn surfaces of friction composite abraded under fine (212 μm) and coarse (425 μm) abrasive particles are shown in Fig. 3.17 (a) and (b). The effect of reinforcement fibers (basalt-recycled aramid) along with hard particles (SiC, Al_2O_3 and iron) has been analyzed. Fig. 3.17 (a) represents the worn surface of composite C1 for 2000 m abrading distance at a 30 N load under the action of silica sand size of 212 μm. The severity and extent of damage to the phenolic matrix surface becomes less because it contains a combination of increased number of hard phases, such as fibers and abrasive particles. As abrasion proceeds, the fibers (marked as M) along with hard particles (marked as H) resist the damaging action of the abrasive particles. Hence, minimum damage was noticed on the abraded surface as shown in Fig. 3.17 (a). The minimal and localized generation of voids (marked as V) is evident in Fig. 3.17 (a). Hence, this surface is relatively smooth with a minimum amount of basalt fiber fragmentation when compared to what is observed in Fig. 3.17 (b).

A good adhesion between the fiber-matrix interfaces is visible in Fig. 3.17 (a) and this obstructs the removal of low modulus phenol formaldehyde matrix during abrasion. This is because its degree of debonding is not enough to cause fiber removal in the friction composite. Also, the extent of plastic deformation and the pulling out of fibers from the matrix is almost negligible, thus lowering the wear of friction composite. Further, the ductility and elongation of soft recycled aramid fiber considerably improve the wear resistance of composites. Thus, this observation lends credence to the contention that the presence of fibers along with hard abrasive particles allows less wear to a matrix during abrasion, which in turn causes less fiber breakage (marked as B). Hence, the resultant worn surface is relatively smooth and resin rich (marked as S) with minimum microcracks and voids, as shown in Fig. 3.17 (a). However, the severity of abrasions such as microcutting and microplowing increases as the size of abrasive particle increases from 212 to 425 μm, resulting in greater damage to the abraded surface of the friction composite when compared to the former one, as shown in Fig. 3.17 (b). Further, Fig. 3.17 (b) illustrates the worn surface of composite C1 for 1500 m abrading distance at a 30 N load under the action of silica sand with size of 425 μm. Due to fatigue action by the repeated abrasion of hard silica sand, a high degree of fiber breakages, fiber debonding, fiber fragmentation (marked as G) and broken fibers (marked as B) are noticed predominantly along the direction of the flow of the abrasives. The microcutting and transverse bending effect of coarse silica sand generates damaged fibers in the surface and makes them pull out (marked as E) from the matrix. The entrapment of sand particles leads to deformation and damages the matrix surface of the composite, resulting in formation of wear debris (marked as D), which can be noticed in Fig. 3.17 (b). Interfacial debonding is caused by poor adhesion between filler, fiber and phenolic matrix, results in an accelerated breakage of matrix interspersed with plowed furrows and displaced fibers corresponding to the direction of wear. Due to the tensile stresses imposed by the abrasives on the surface, the fiber torn (marked as T) or peeling off from the matrix is another extended feature. The low modulus phenol formaldehyde matrix is severely damaged and removed quickly with the development of various types of cracks on the surface like fatigue cracks (marked as C) due to the plowing action of abrasives with hard debris (marked as D) particles.

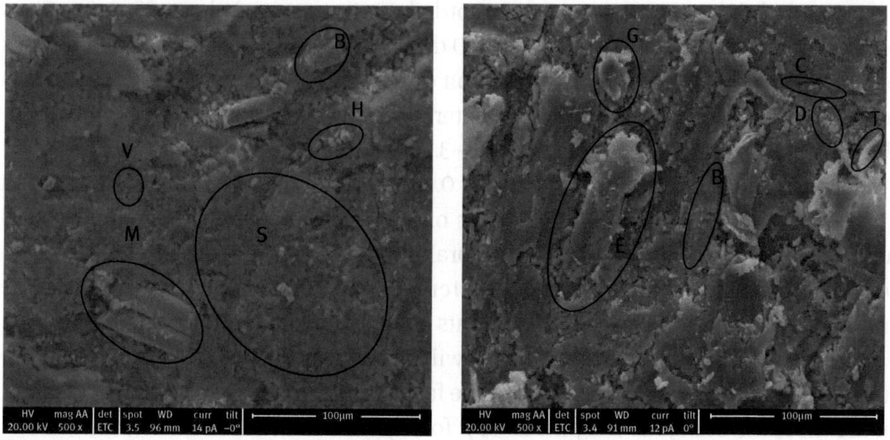

(a) (b)

Fig. 3.17: (a) Worn surface morphology of composite C1 for 2000 m abrading distance at a 30 N load under the action of silica sand size of 212 μm (b) Worn surface morphology of composite C1 for 1500 m abrading distance at a 30 N load under the action of silica sand size of 425 μm.

3.3.3 Fade and recovery behavior of basalt-recycled aramid fiber reinforced friction composites

Fade is a detrimental behavior which occurs due to thermal degradation of organic polymer matrix, formation of load carrying friction film at the friction interface and loss of mechanical properties of the friction composites [11]. The friction composites should possess higher and stable μ and lower wear under various operating conditions [30]. From the tribological study of friction composites, the composites C1, C2, and C3 show better performance towards dry sliding conditions. Hence, it was decided to study the influence of fiber reinforcement and effect of thermal stability on fade-μ and recovery-μ behavior of hybrid friction composites C1, C2, and C3 and this is explained in the following sections.

3.3.3.1 Fade and recovery cycles of hybrid friction composites

The fade and recovery responses of composites C1, C2, and C3 at a constant speed of 411 rpm, constant interval of 10 s and constant load of 617 N are shown in Fig. 3.18–3.20. The figure incorporates the performance μ (average μ over the whole braking), the fade μ, and the recovery μ. This investigation evaluates the fade and recovery behavior of friction composites by variation in μ with a rise in disc temperature during fade and recovery stage of the friction test. During the friction test, the fade and recovery μ and disc temperature were recorded continuously. At least 20 data points were recorded at each stage. Fig. 3.18–3.20 describes the fade and recovery behaviors of hybrid friction composites as a function of temperature. Fade designates the loss in braking effectiveness of a friction material at elevated temperature, typically in the range of 250 to 400 °C. From Fig. 3.18–3.20 it can be observed that the developed composites showed the μ in the range of 0.26 to 0.55, which is within the industrial applications range. Generally, the values of μ of friction composite lie between 0.3 and 0.6 [31]. The variations of μ during braking depend on changes in the real area of contacts at the friction interface, the strength of the phenolic binder resin and the frictional characteristics of the constituents at elevated temperatures.

During braking, the temperature rise will be between 50–350 °C and this temperature zone is very crucial to determining the fade characteristics of friction composites. In the first fade cycle, the magnitude of μ for composite C1 increases slightly with an increase in temperature up to 200 °C and then decreases and nearly reaches the initial position of the test. The increase in μ with increasing temperature is mainly due to the formation of primary plateaus. The primary plateaus are formed by broken fibers and other hard inclusions which are embedded on the surface of the composites. However, the composites C2 and C3 follow the trend of decreasing μ with an increase in drum temperature and are subject to more frictional undulations. On the other hand, during the second fade cycle, the overall magnitude of μ dropped down prominently around 300 °C and remained constant until 350 °C. At higher temperature, the low molecular

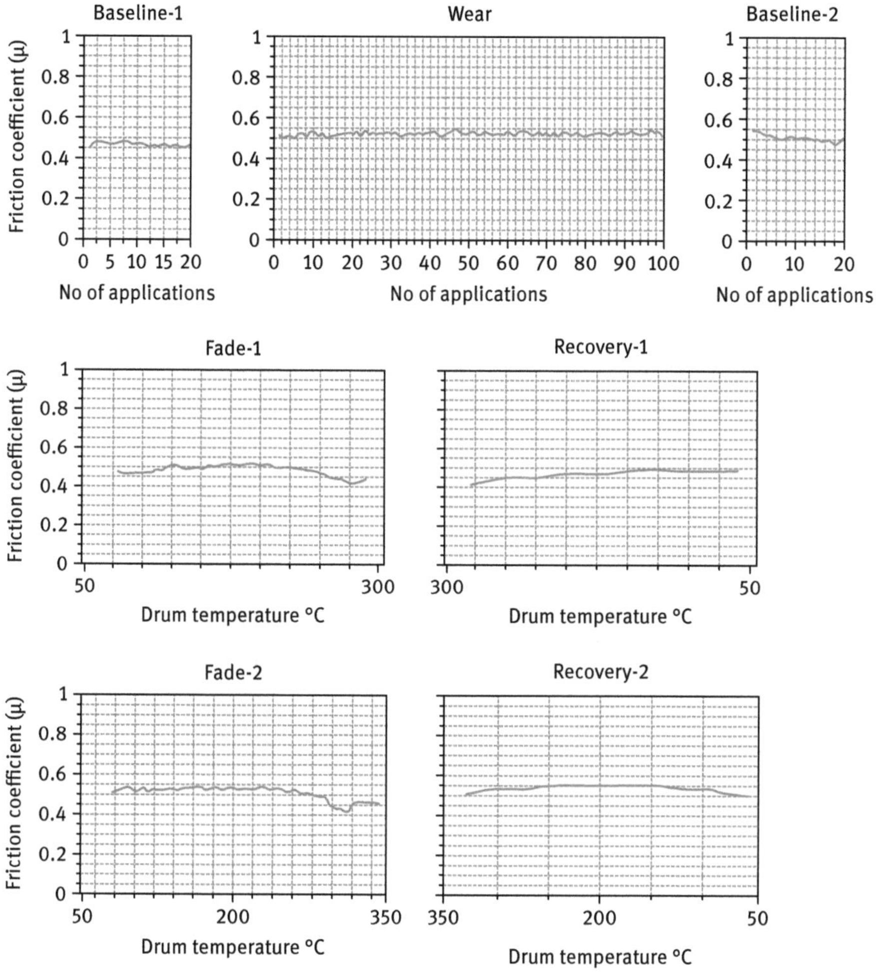

Fig. 3.18: Fade and recovery cycles of friction composite C1.

weight of phenol formaldehyde resin in the friction composites gets burnt off rapidly. Hence, phenol formaldehyde resin starts to melt and act as a lubricant between the friction composite and the rotor drum and causes the brake to slip. Also, the amount of glazed layer of the phenol formaldehyde resin present on the surface leads to a reduction in the restoration of the μ. The decrease in μ was attributed to the increase in sliding temperature between friction composite and brake drum. On the contrary, during the recovery test, the μ of composite C1 increases with decreasing temperature and quickly recovered to its original magnitude. However, the recovery behavior of composites C2 and C3 is very poor due to their lower thermal stability behavior. The revival of μ during the recovery cycle depends on the formation of types of friction film (plateaus) on the surface after the completion of the fade cycle.

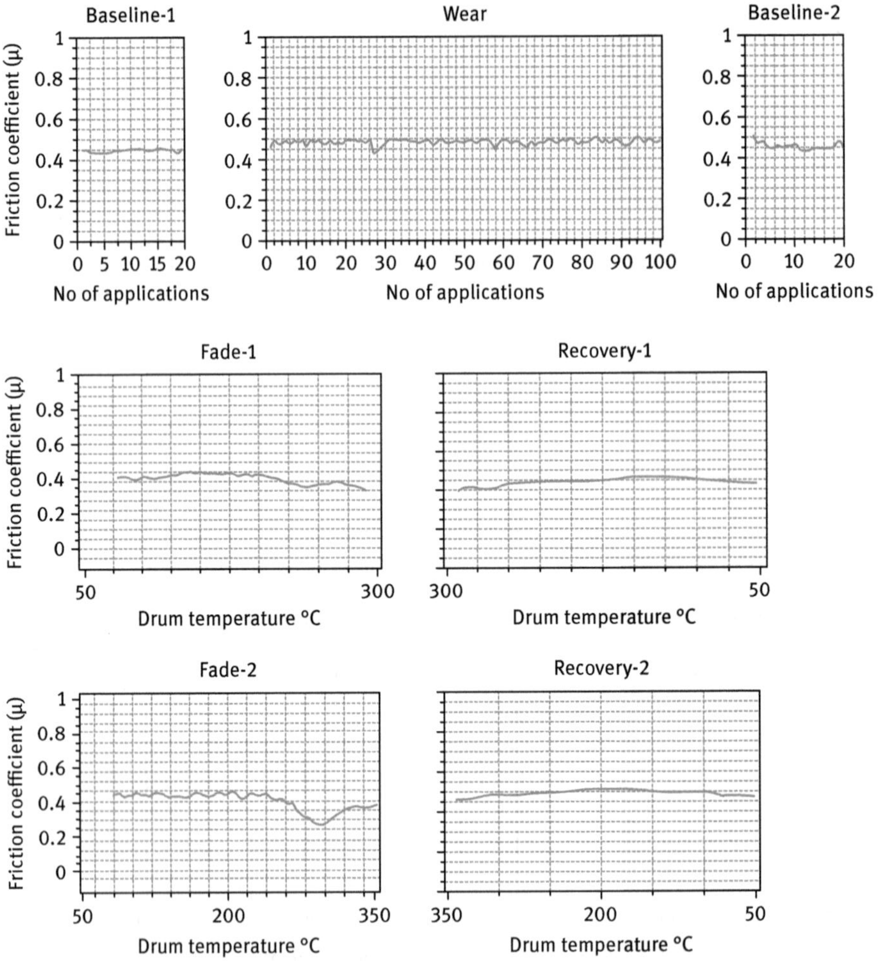

Fig. 3.19: Fade and recovery cycles of friction composite C2.

From Fig. 3.18–3.20 it can be confirmed that, during the fade test, the μ of composites C1, C2 and C3 decreases with an increase in drum temperature, whereas in the recovery test, the μ of composites increases with decreasing temperature. The reason for this is that during braking the frictional heat is generated at the friction interface of the composites and then the interface surfaces and the disc are allowed to cool down on release of the brake which causes a shear-induced thermomechanical stressed state in the friction composites and relaxation of the friction surface. The frictional surface relaxation thus developed reduces the deformation of the heterogeneous friction film of the friction composites. This reduction in friction film deformability imparts brittleness in the friction film, which easily ruptures or disintegrates the heterogeneous the friction layer. Further, the disintegration of the heterogeneous friction film forms

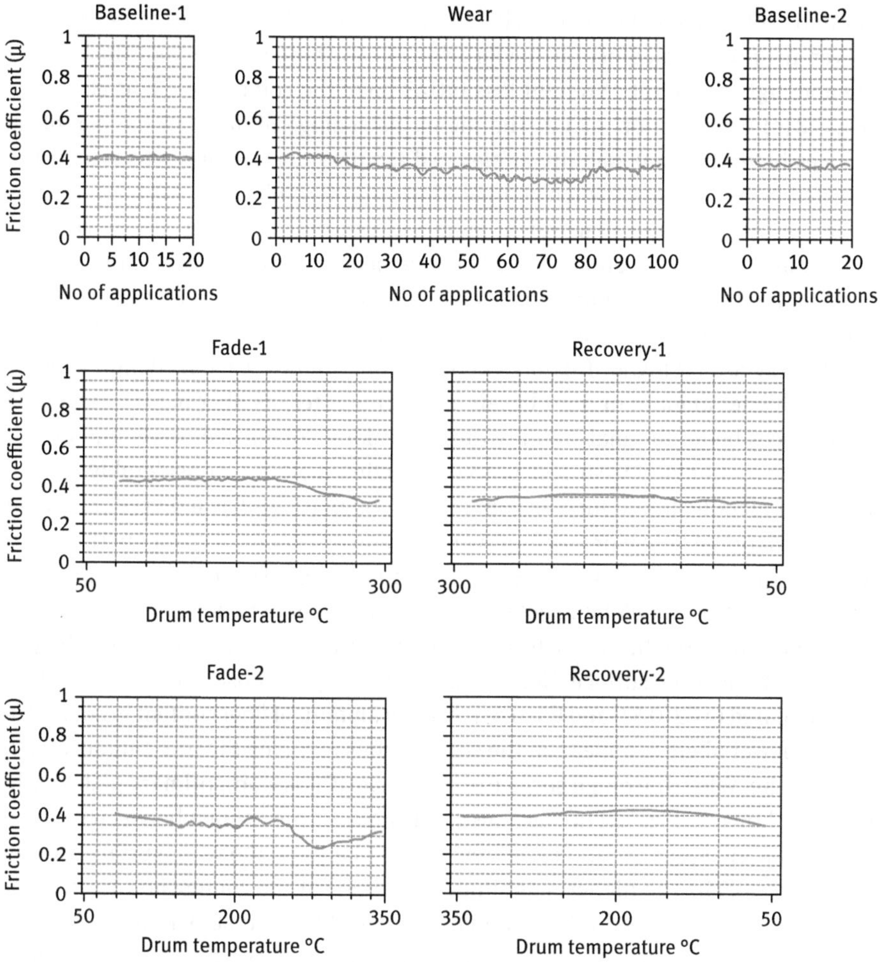

Fig. 3.20: Fade and recovery cycles of friction composite C3.

wear debris at the friction interface by interfacial compositional domain incompatibility, thermal crazes and also generates a new surface before it is subjected to a further braking cycle [11]. Therefore, the degree of recovery behavior of composites is enhanced when the wear debris particles are harder or the adhesive debris particles are higher. But, however, this wear debris gets entrapped between the friction surfaces and forms the third bodies, resulting in three-body abrasion (TBA) in the friction composites. Hence, the fade behavior of composites C2 and C3 is boosted and reduces the frictional performance of the friction composites when compared to composite C1.

The initial μ instability from the baseline plots is not unusual and normally disappears after a certain number of brake applications in the real braking situation. During the second fade test, the μ of the composites goes to the lowest level of 0.26

at around 300 °C in composite C3. The μ of composite C3 was found to be very low, not even near to the mediocre behavior which is attributed to the poor thermal stability and higher amount of weight loss (40.43 %) of the composite C3, as confirmed in Fig. 3.8 (c) from Section 3.3.1.2. The higher percentage of fibers in the composite C1 causes better thermal stability at elevated temperatures. Hence, less weight loss is attested in composite C1 when compared to other formulations. Also, the thermal mobility of the low molecular weight is higher, resulting in easier plastic deformation, leading to a larger contact area. The fade behavior at elevated temperatures is related to the thermal decomposition of ingredients and is followed by a subsequent destruction of contact areas at the sliding interface. So the higher thermal stability lined the way for the higher friction stability in the composite C1. A friction composite with a higher amount of hard basalt fiber content resists fiber matrix debonding and enhances better thermal stability during braking. Fibers like basalt and aramid improve the friction and wear resistances of friction materials considerably [10, 32]. Addition of metallic fillers such as copper and iron particles has a diverse role in altering the frictional behavior and improves the fade resistance, recovery performance and counterface friendliness of the friction composite [33]. The μ of the composites is mostly governed by the plowing action of the reinforcement fiber present in the composites [34]. Thus, the hard basalt fiber enhances the plowing action of the friction composite on the drum surface. Hence a higher μ (0.55) was observed in composite C1 when compared to C2 (0.52) and C3 (0.45). A lower amount of fiber content in composite C3 causes fewer fibers to appear on the surface, thus reducing the true contact area of the composite. When the rough surface of composite (C1) is worn, primary plateaus are formed as shown in Fig. 3.21 (a), as a result the area of true contact between composite and drum increases. An increased real contact between the worn surfaces results in increased μ [35]. The higher magnitude of μ (0.55) of the composite C1 can be attributed to the unique head-tail morphology of basalt fiber, as shown in Fig. 3.2 (a) in Section 3.2.1.1. During braking, the heads in basalt fiber act as hard abrasive particles in the composites, resulting in an increase in true contact between the friction pair and leading to two-body abrasive wear grooving mechanism in the friction composites [11]. This has turned into successfully imparting the higher μ in composite C1 when compared to other formulations.

In the friction composites, basalt fiber exhibits high potential and receives a lot of attention due to its high temperature and abrasion resistance. Incorporation of soft aramid fiber in the friction formulations also enhances the μ of the composites by forming the viscous glassy layer on the area of contact of the counterpart [36]. The aramid fiber provides stable and higher friction stability and better thermal stability to the friction surfaces and also boosts frictional performances in the composite C1. Also the reinforcement fibers along with hard abrasives such as Fe, Cu, Al_2O_3, and SiC in the friction formulations increase the true contact area between the friction surfaces, which leads to higher μ in the composite C1. The molykote in the friction composites smeared from the friction surface during sliding acts as a lubricant and reduces the

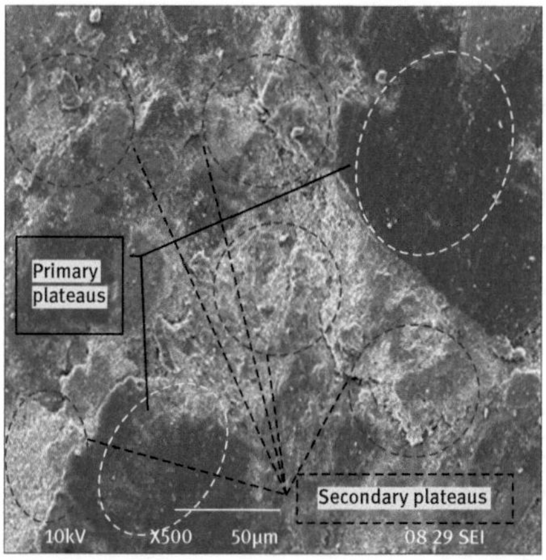

(a)

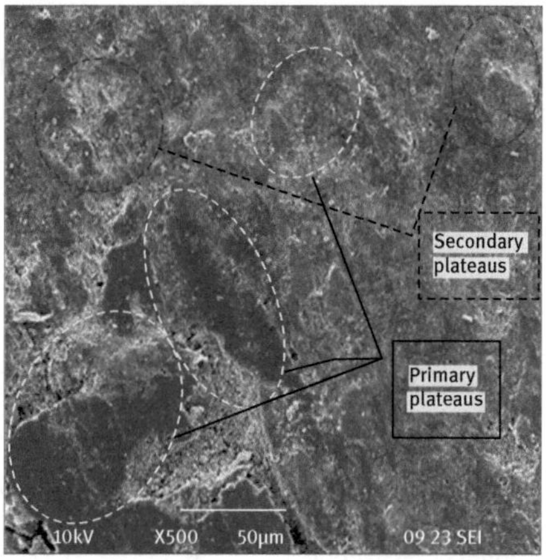

(b)

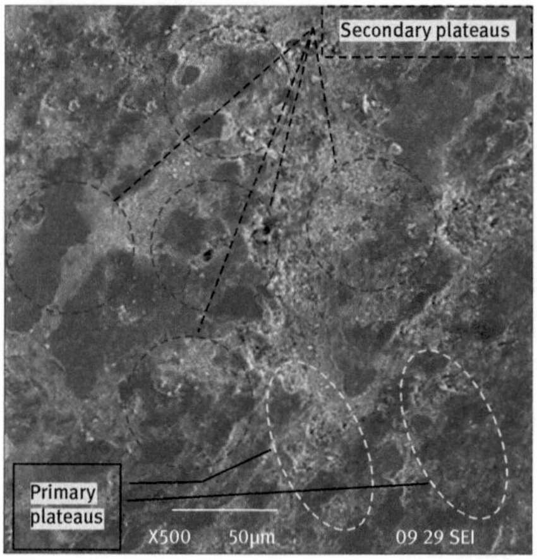

(c)

Fig. 3.21: SEM images of friction composite samples (a) C1, (b) C2, and (c) C3 respectively.

frictional fluctuations of the composites. Incorporation of cashew friction dust particles into the friction material offer adequate elasticity resulting in stabilized frictional characteristics. This is because the contact area of friction surfaces increases through the elastic deformation on engaging with the counterface of the friction composite at lower temperatures. Hence, higher μ is maintained in the case of composite C1 throughout the process when compared to other formulations. The various frictional responses of the friction composites are shown in Tab. 3.18. At high temperatures, the organic constituents in the friction composites release gas as a result of thermal degradation or pyrolysis. This gas may be trapped between the sliding interface and an opposing force to the applied load results in reducing the frictional force. Therefore, the fade behavior of friction composites at elevated temperature is due to the thermal decomposition of ingredients and is followed by a consequent damage of contact areas at the sliding interface. Also, high strength and high modulus fibers in the composites resist thermal degradation by absorbing the major share of stress and dissipate a lower amount of energy as a friction output. It has been reported that the fade behavior of the friction material could be dominated by a decrease in mechanical strength, thermal degradation of the organic matrix and the formation of load-carrying friction film causing an effective increase in the real contact area and consequently a reduction in applied pressure on the friction material composites [37, 38].

Tab. 3.18: IS 2741 Part 4 (SAE J661) frictional response of the friction composites.

Performance attributes	C1	C2	C3
Hot friction or fade μ	0.51	0.50	0.48
Normal or recovery μ	0.55	0.52	0.44
First fade μ (%) [a]	10.42	15.91	28.21
First recovery μ (%) [b]	116.67	105.13	93.94
Second fade μ (%)[c]	7.55	10.42	20.93
Second recovery μ (%) [d]	96.00	102.17	89.19
Wear loss (%)	8.08	8.78	8.95

During the friction tests, the coefficient of friction and the drum temperature were recorded continuously. At least 20 data points were recorded for each stage. Two types of friction values, i.e., normal friction coefficient (average of four points in the second fade curve located at 93, 121, 149, and 205 °C) and hot friction coefficient (average of 10 points located at 206 and 149 °C on the first recovery; 233, 261, 127.2, 142.8, 158.3, and 173.9 °C on the second fade; 260.7, 205, and 149 °C on the second recovery) are calculated. The percentage of first fade μ, first recovery μ, second fade μ, and second recovery μ are calculated which are defined as follows.

Calculations

a $$\text{first fade } \mu\ (\%) = \left\{\frac{\text{initial } \mu - \text{final } \mu}{\text{initial } \mu}\right\} \times 100 \text{ (during first fade cycle)}$$

b $$\text{first recovery } \mu\ (\%) = \left\{\frac{\text{final } \mu}{\text{initial } \mu}\right\} \times 100 \quad \text{(during first recovery cycle)}$$

c $$\text{second fade } \mu\ (\%) = \left\{\frac{\text{initial } \mu - \text{final } \mu}{\text{initial } \mu}\right\} \times 100 \text{ (during 2nd fade cycle)}$$

d $$\text{second recovery } \mu\ (\%) = \left\{\frac{\text{final } \mu}{\text{initial } \mu}\right\} \times 100 \quad \text{(during 2nd recovery cycle)}$$

μ_{max}: the highest coefficient of friction during fade and recovery test
μ_{min}: the lowest coefficient of friction during fade and recovery test

Magnitude of fade μ and % of fade for all the composites was in the following order (Tab. 3.18):
Fade μ: C1 (0.51) > C2 (0.50) >C3 (0.48)
First fade μ (%): C1 (10.42) < C2 (15.91) < C3 (28.21)
Second fade μ (%): C1 (7.55) < C2 (10.42) < C3 (20.93)

Magnitude of fade μ and % of recovery for all the composites was in the following order (Tab. 3.18)
Recovery μ: C1 (0.55) > C2 (0.52) > C3 (0.44)
First recovery μ (%): C1 (116.67) > C2 (105.13) >C3 (93.94)
Second recovery μ (%): C1 (96.00) < C2 (102.17) > C3 (89.19)

The higher damage to the friction surface in composite C3 could be due to interfacial thermochemical changes in the friction composite during braking. At elevated temperatures, the surface energy at the friction surfaces is reduced and generates a related temperature friction surfaces between friction composite and counterface drum resulting in decreased adhesive-cohesive forces between the interfacial friction zones [39]. This causes the mating surfaces of the friction couple to slip over each other without difficulty and resembles the thermodynamical nature of the friction film. This friction film imparts the related rheological responses to counter the thermoelastic instability (TEI) at the friction interface and deforms the contact zone while braking [40, 41]. Hence, the μ of the friction composite C3 was lower and this contributes to the fade behavior when compared to composites C2 and C3.

3.3.3.2 Wear behavior of friction composites and brake drum

The tribological behaviors are basically evaluated by the nature of interactions between the two friction surfaces that are in contact. The nature and level of interactions between the two surfaces are dependent on the properties of the surfaces and the ingredients of composites [34]. In the friction material composites, the phenolic resin, fibers, and other organic friction modifiers play crucial roles in the wear of the friction materials which occurs due to temperature transformation of these ingredients [42]. In general, softening and charring of phenol formaldehyde matrix increase with an increase in drum temperature, which is a common phenomenon, and this temperature rise causes fiber-matrix debonding at the fiber-matrix interface resulting in detachment of the polymer matrix, and cracking of the deteriorated surface layer [43]. Basalt fiber exhibits high hardness, excellent fiber-matrix adhesion and abrasion resistance to friction materials. Aramid fiber is soft and improves the wear resistance of composites due to its higher strength and characteristic properties of fibrillation and elongation. Due to its fibrillation property, it absorbs a higher amount of stress and reduces the residual stress acting at the interface in the case of the abrasive wear [29].

In the hybrid friction composite, the addition of basalt fiber (20 wt%) in composite C1 exhibits a lower wear rate when compared to all other composites. The composite with 25 wt% of fiber (C1, basalt + recycled aramid) is more resistant to fiber-matrix debonding due to the matrix keeping its integrity at high temperature. However, composite C3 loses it surface integrity at high temperature due to its fiber-matrix debonding behavior. At high temperature, the thermal degradation and softening of the phenol formaldehyde resin strongly influence the tribological behavior of the composite. Removal of the fiber due to thermal degradation is not the only cause of the higher wear. Also, thermal softening and charring effects on the matrix result in debonding of the fibers and enhance fiber pull-out from the matrix, resulting in loss of structural integrity. A lesser amount of basalt fiber in the composite C3 results in increased debonding of reinforcing fibers and fracture of matrix. These deboned fibers can cause an increase in wear rate of the composites by damaging the counterface film. The thermal softening of the matrix reduces the hardness of the composites and thus increases the wear rate of the composite C3. Hence, a higher wear rate was observed in composite C3. The wear loss of the composites follows the trend of C1 (8.08 %) < C2 (8.78 %) < C3 (8.95 %) as listed in Tab. 3.19.

Tab. 3.19: Wear loss due to weight loss of hybrid friction composites.

S. No.	Composite	Weight loss (%) from TGA	Wear loss (%) from chase friction test
1	C1	33.32	8.08
2	C2	38.57	8.78
3	C3	40.43	8.95

3.3.3.3 Worn surfaces morphology of hybrid friction formulations

In the SEM analysis, primary plateaus are denoted by dark colored patches which in turn have good load carrying capacity, fade resistance ability and reduced wear. The secondary plateaus are denoted by the shiny colored patches that are formed by the back transfer of organic material such as resin, which is either due to adhesive wear or due to the hot spots formed in the rotor [44]. The load carrying capacity of secondary plateaus is poor due to charred materials, which in turn reduce friction and induce wear. The secondary plateau formation was high in C3 followed by C2 where even more surface cracks are formed, but C1 showed more primary plateau with less secondary plateau and no surface crack formation. This reduced fade behavior can be attributed to its good thermal stability and higher thermal conductivity. The composite C3 was rougher and covered with secondary plateaus supporting the higher extents of third-body abrasion in composites which is discussed in Section 3.3.2. Also, in composite C3, thermal cracks are evident due to its softening and thermal degradation of the matrix, as evident in Fig. 3.21 (c). The fiber pull-out due to thermal penetration, thermal softening and charring is greater in composite C3 when compared to composites C1 and C2. Generation of microcracks in composite C3 facilitates more wear as compared to the other composites. A higher amount of basalt fiber in the composite C1 worked in a synergistic manner – showing a minimal amount of surface damage other than the formation of secondary plateaus. The basalt-recycled aramid fiber combination was antagonistic to the wear debris of the composite C1, while surface damage due to broken fibers, hard abrasive debris, secondary plateaus and low-intensity surface cracks are evident in the composite C3.

3.4 Conclusion

The density of hybrid friction composites decreased linearly with an increase in wt% of fiber content whereas the heat and water swelling of the friction composites increase with the addition of reinforcement fibers. Further, acetone extraction and the loss of weight on ignition at 800 °C of the friction composites decrease with an increase in wt% of fiber content.

The mechanical properties such as hardness and strength of friction composites increased with an increase in wt% of reinforcement fiber. The composite C1 with higher amount of fiber (25 wt%) is 1.2 times harder than composite with a lower fiber content (C5, 5 wt%). The shot being hard in basalt fiber exhibits a higher degree of resistance to indentation and restricts plastic deformation of the hybrid friction composites.

The composite with an increased amount of reinforcement fiber (C1) exhibits the highest tensile strength (12.85 MPa) and elongation (1.15 mm), better compressive strength (66.96 MPa) by neglecting the fiber crushing failures and possesses superior flexural strength (50.09 MPa). The impact energy of friction composites varies from 0.151 to 0.451 Joules with increasing fiber content from 5 wt% (C5) to 25 wt% (C1).

Addition of basalt fiber improves the thermal stability of friction composites. The friction composites with a higher amount of fiber content (C1, 25 wt%) show better thermal stability and minimum weight loss which follows the trend of C1 (33.32 %) < C2 (38.57 %) < C3 (40.43 %) < C4 (40.82 %) < C5 (47.23 %).

The friction composite with an increased amount of fiber (basalt + recycled aramid) content (C1, 25 wt%) shows better abrasive wear resistance when compared to composites reinforced with decreased amounts of basalt fiber content (C5, 5 wt%). The severity of damage to the friction composite surface is greater under the action of coarse (425 μm) sand than that of fine (212 μm) sand particles. The coarse abrasives are harder and remove the material at a faster rate when compared to fine abrasives.

From the Grey-based Taguchi approach, the optimum factor combination for multiple responses (specific wear rate, wear coefficient and hardness) were evaluated from the values of GRG and in the present experimental study it is $A_1L_1F_3D_3$, i.e., silica sand abrasive size of 212 µm (A_1), applied load of 30 N (L_1), composite with 25 wt% fiber content (F_3) and abrading distance of 2000 m (D_3) under three-body abrasion condition. The ANOVA test confirmed that the wt% of fiber content (58.85 %) was found to be the most effective factor which affects the abrasive wear behavior of composites followed by abrasive particle size (21.55 %), abrading distance (9.07 %) and applied load (3.77 %).

The fade and recovery test confirmed that the frictional performances are significantly influenced by the amount of fiber reinforcement in the composites. The basalt-recycled aramid fiber combination in composite C1 (25 wt%) exhibits a better fade resistance and recovery behavior when compared to composites C2 and C3. The basalt-recycled aramid fiber combination was found to be antagonistic to the wear debris in composite C1 and the surface damage due to broken fibers and hard abrasive debris.

The higher fiber (25 wt%) containing friction composite (C1) resists thermal degradation and softening of the phenol formaldehyde resin due to its higher thermal stability behavior resulting in an increased friction coefficient and decreased specific wear rate. The composite C1 was found to exhibit more primary plateaus and fewer secondary plateaus and no evidence for formation of surface cracks. The secondary plateau formation was high in C3 followed by C2 where even more surface cracks are formed. Increased thermal cracks and fiber pull-out were found in composite C3 due its higher thermal degradation and fiber-matrix debonding behavior of the composite.

References

[1] Filip P, Weiss Z, Rafaja, D. On friction layer formation in polymer matrix composite materials for brake applications. Wear. 2002;252(3–4):189–198.

[2] Bijwe J. Composites as friction materials: Recent developments in non-asbestos fiber reinforced friction materials – A review. Polym Compos. 1997;18(3):378–396.

[3] Mukesh Kumar, Satapathy BK, Patnaik A, Kolluri DK, Tomar BS. Hybrid composite friction materials reinforced with combination of potassium titanate whiskers and aramid fibre; Assessment of fade and recovery performance. Tribo Intern. 2011;44(4):359–367.

[4] Manoharan S, Suresha B, Bharath PB, Ramadoss G. Investigations on three-body abrasive wear behaviour of composite brake pad material. Plas Polym Tech. 2014;3:10–18.

[5] Rhee SK, Jacko MG, Tsang PHS. The role of friction film in friction, wear and noise of automotive brakes. Wear. 1991;146(1):89–97.

[6] Trezona RI, Allsopp DN, Hutchings IM. Transition between two-body and three-body abrasive wear: Influence of test conditions in micro scale abrasive wear test. Wear. 1999;225–229:205–214.

[7] Woldman M, Heide EVD, Schipper DJ, Tinga T, Masen, MA. Investigating the influence of sand particle properties on abrasive wear behavior. Wear. 2012;294–295:419–426.

[8] Chand N, Naik A, Neogi, S. Three-body abrasive wear of short glass fibre polyester composite. Wear. 2000;242:38–46.

[9] Cenna AA, Allen S, Page NW, Dastoor P. Modelling the three-body abrasive wear of UHMWPE particle reinforced composites. Wear. 2000;254(5–6):581–588.

[10] Öztürk B, Arslan F, Öztürk, S. Hot wear properties of ceramic and basalt fiber reinforced hybrid friction materials. Tribo Intern. 2007;40(1):37–48.

[11] Satapathy BK, Bijwe J. Fade and recovery behaviour of non-asbestos organic (NAO) composite friction materials based on combinations of rock fibers and organic fibres. J Reinf Plast Comp. 2005;24:563–577.

[12] Subbaya KM, Rajendra N, Varadarajan YS, Suresha, B. Multiple response optimization of three-body abrasive wear behaviour of graphite filled carbon-epoxy composites using Grey-based Taguchi approach. J Min Mat Charac Eng. 2012;11:876–884.

[13] Hao LC, Yu WD. Comparison of morphological structure and thermal properties of basalt fiber and glass fiber. J Xi'an Poly Univ. 2009;23(2):327 – 332.

[14] Botev M, Betchev M, Bikiaris HD, Panayiotou, C. Mechanical properties and visco elastic behaviour of basalt fiber-reinforced polypropylene. J App Poly Sci. 1999;74(3):523–531.

[15] Kato T, Magario A. The wear of aramid fiber reinforced brake pads: The role of aramid fibers. Tribo Trans. 1994;37(3):559–565.

[16] Bijwe J, Indumathi J, Ghosh AK. On the abrasive wear behaviour of fabric-reinforced polyetherimide composites. Wear. 2002;253(7–8):768–777.

[17] Manoharan S, Suresha B, Ramadoss G Bharath PB. Effect of short fiber reinforcement on mechanical properties of hybrid phenolic composites. J Mat. 2014;2014:1–9.

[18] Manoharan S, Ramadoss G, Suresha B, Vijay R. Influence of fiber reinforcement and abrasive particle size on three-body abrasive wear of hybrid friction composites. App Mech Mat. 2015;766–767:156–161.

[19] Zum Gahr KH. Wear by hard particles. Tribo Intern. 1998;31:587–596.

[20] Ross PJ. Taguchi techniques for quality engineering. New York: McGraw-Hill; 1988.

[21] Basawarajappa S, Chandramohan G, Paulo Davim J. Application of Taguchi techniques to study dry sliding wear behaviour of metal matrix composites. Mat Des. 2007;28(4):1393–1398.

[22] Kuo Y, Yang T, Haung GW. The use of Grey-based Taguchi method for optimizing multi-response simulation problems. Eng Optim. 2008;4(6):517–528.

[23] Manoharan S, Ramadoss G, Suresha B, Vijay R. Effect of fiber reinforcement on tribological performances of hybrid friction composites. J Balk Tribol Asso. 2015;21(3): 525–538.

[24] Talegaonkar RP, Gopinath K. Influence of alumina fiber content on properties of non-asbestos organic brake friction material. J Reinf Plast Compo. 2008;1–14.

[25] Basawarajappa S, Ajith GJ, Arun KV, Praveen Kumar A, Prasanna kumar M. Three-body abrasive wear behaviour of polymer matrix composites filled with SiC particles. Poly Plast Techn Eng. 2009;49(1):8–12.

[26] Rabinowicz E, Dunn LA, Russell PG. A study of abrasive wear under three-body conditions. Wear. 1961;4(5):345–355.

[27] Suresha B, Chandramohan G, Siddaramaiah, Sampathkumaran P Seetharamu S. Three-body abrasive wear behaviour of carbon and glass fiber reinforced epoxy composites. Mat Sci Eng: A. 2007;443(1–2):285–291.

[28] Ramesh BN, Suresha B. Optimization of tribological parameters in abrasive wear mode of carbon-epoxy hybrid composites. Mat Des. 2014;59:38–49.

[29] Satapathy BK, Bijwe J. Analysis of simultaneous influences of operating variables on abrasive wear of phenolic composites. Wear. 2002;253(7–8):787–794.

[30] Briscoe BJ, Ramirez I, Tweedle PJ. Friction of aramid fiber composites. In: Proceedings of the International Conference on Disc Brakes for Commercial Vehicles, The Institution of Mech Eng, London; 1998;1:15–29.

[31] Blau PJ, McLaughlin JC. Effects of water films and sliding speed on the frictional behaviour of truck disc brake materials. Tribo Intern. 2003;36(10):709–715.

[32] Sinha SK, Biswas SK. Effect of sliding speed on friction and wear of unidirectional aramid fibre/phenolic composite. J Mat Sci. 1995;30:2430–2437.
[33] Mukesh Kumar, Bijwe J. NAO friction materials with various metal powders, tribological evaluation on full-scale inertia dynamometer. Wear. 2010;269(11–12):826–837.
[34] Crosa G, Baumvol IJR. Tribology of polymer composites used as frictional materials. Advances in Composite Tribology. Netherlands: Elsevier; 1993. p. 582–626.
[35] Eriksson M, Jacobson S. Tribological surface of organic brake pads. Tribo Intern. 2002;33:817–827.
[36] Horrocks AR, Eichhorn H, Schwaenke H, Saville N, Thomas C. Thermally resistant fibres, high performance fibres. Cambridge: CRC Press; 2001. Chapter 9, 281–324.
[37] Herring JM. Mechanisms of brake fade in organic brake pad linings. Soc Auto Eng. 1967; 670146.
[38] Wirth A, Eggleston D, Whitaker R. A fundamental tribo-chemical study of the third body layer formed during automotive friction braking. Wear. 1994;179:75–81.
[39] Williams JA. The friction of solids. In: Engineering Tribology. New York: Oxford University Press; 1994. Chap. 4, 132–164.
[40] Lee K, Barber JR. Frictionally excited thermo elastic instability in automotive disk brakes. ASME J Tribo. 1993;115:607–614.
[41] Dow TA. Thermo elastic effects in brakes. Wear. 1980;59:213–221.
[42] Shin MW, Cho KH, Lee WK, Jang H. Tribological characteristics of binder resins for brake friction materials at elevated temperatures. Tribology Letters. 2010;38(2):161–168.
[43] Tanaka K, Ueda S, Noguchi N. Fundamental studies on the brake friction of resin-based friction materials. Wear. 1973;23(3):349–365.
[44] Choa MH, Cho KH, Kim SJ, Kim DH, Jang H. The role of transfer layers on friction characteristics in the sliding interface between friction materials against grey iron brake disks. Tribo Lett. 2005;20(2):101–108.

J. Babu, Jose Philip, and J. Paulo Davim

4 Machining defects in green composites

Notation

$A_d = D_{MAR}$ Delamination area in the vicinity of the drilled hole
A_e Area of the envelope of damage zone which includes area of the hole
A_H Heavy damage area
A_L Light damage area
A_{max} Area corresponding to D_{max}
A_M Medium damage area
$A_0 = A_{nom} = A_{AVG}$ Drilled area with diameter D; which is the nominal hole area
BFRC Banan fiber reinforced polymer composites
CFRC Carbon fiber reinforced polymer composites
$D = D_0$ Nominal diameter of drilled hole
D_e Equivalent diameter
$D_{ea} = D_{re}$ Equivalent diameter of a circle whose area is the same as A_e
D_{ep} Equivalent diameter of a circle whose perimeter is the same as P_e
DF = Di Damage factor
D_{eed} Effective equivalent diameter (average value of D_{ea} and D_{ep})
D_{min} Diameter of the minimum enclosing area
D_{max} Maximum diameter of delamination
D_{RAT} Damage ratio
f Shape's circularity
F Feed rate
F_a Two-dimensional delamination factor
F_d Conventional delamination factor
F_{da} Adjusted delamination factor
$F_{d_{min}}$ Minimum delamination Factor
F_{DR} Refined delamination factor
F_{ed} Equivalent delamination factor
F_{red} Refined equivalent delamination factor
FEED Effective equivalent delamination factor
FML Fiber-metal reinforced composite laminate
FRPC Fiber reinforced polymer composites
GFRC Glass fiber reinforced polymer composites
HSS High speed steel
HRPC Hemp fiber reinforced polymer composites
P_e Perimeter of the envelope of damage zone
NCM Non-conventional machining processes
NFRCs Natural fiber reinforced composites
NFRPs Natural fiber reinforced plastics
PEEK Polyetheretherketone
PLA Polylactic acid
R_a Surface roughness
R_{max} Maximum radius of damaged area
R_{nom} Nominal radius of the drill

https://doi.org/10.1515/9783110435788-005

R_z Average maximum height
S Spindle speed
SC Silicon carbide
SEM Scanning electron microscope
TiN Titanium nitride
WPC Wood plastic composite
W_{max} Width of maximum damage in mm
W Nominal width of the cut in mm.

4.1 Introduction

Engineering composites are finding wide use in many fields which include defense, transportation and power generation sectors [1–3]. In recent years, environment concerns are driving the use of natural fibers in place of synthetic fibers in fiber reinforced composites. Natural fibers, in addition to being renewable, have low cost/weight, are good thermal insulators and provide high specific strength [4]. Natural fibers are available in nature abundantly and hence can replace expensive nonrenewable, and hazardous synthetic fibers [5, 6].

Manufacturing of products with natural fiber reinforced composites (NFRCs) can be classified as primary and secondary manufacturing. Most of the products with these NFRCs are made to near net shape with primary manufacturing processes like hand lay-up, vacuum bag forming, resin transfer molding, compression molding and pultrusion, etc. [7]. Many defects can occur in the primary manufacturing, such as resin rich/starved area, voids, matrix imperfection cracks, and blisters. These defects are not discussed here.

Although NFRC products are made to near net shape, they may still need some secondary manufacturing (machining) to meet dimensional and/or assembly requirements. The secondary operations usually carried out on NFRC products are: drilling, milling and turning. This chapter presents a discussion on these machining defects in the green composites.

4.2 Defects in drilling of NFRCs

Drilling is a very common secondary machining operation for facilitating assembly of parts made of NFRCs [8, 9]. Defect-free drilling is crucial since drilling is a final operation as no reaming is usually carried out on composites [10]. There are several methods to make holes in composite materials, of which conventional drilling is the most widely used method [11]. Drilling defects on composite materials can be listed as: peel-up delamination, push-out delamination, geometric defects, and thermal damages [12]. Fig. 4.1 represents the factors that influence hole quality in drilling of NFRCs. The most important among them are listed below [13]:

1. the nature and the mechanical characteristics of the fibers used in the composites
2. the fiber orientation
3. fiber/matrix bonding and compatibility
4. machining conditions and parameters
5. machining mode
6. tool material and tool wear mechanism
7. tool geometry

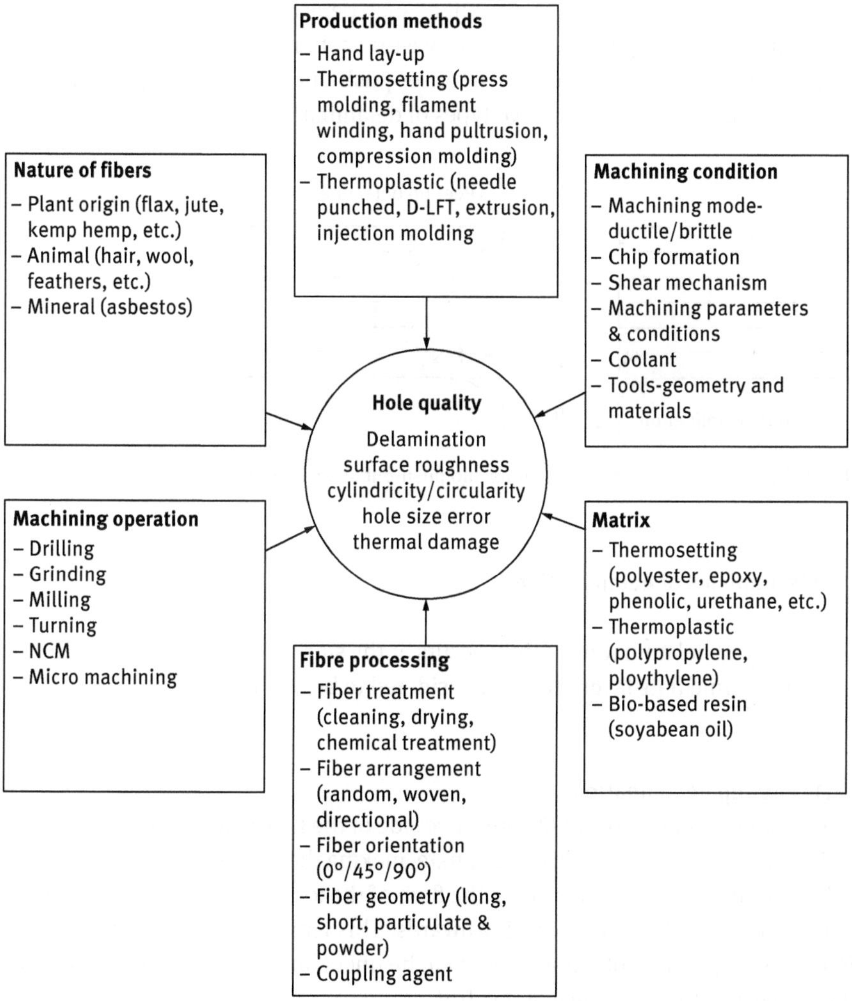

Fig. 4.1: Factors affecting the hole quality in machining of NFRCs.

4.2.1 Delamination in NFRCs

Delamination is defined as separation of plies from each other in case of layered composites. It occurs at the entrance of the hole when the drill bit comes into contact with the work piece and at the exit of the hole when the drill bit reaches the bottom of the work piece. These are known as peel-up deamination and push-out delamination respectively.

Delamination is a major form of damage encountered in drilling [14]. Rejection rates of up to 60 % are reported as a result of such defects indicating its influence on the performance of the parts [15]. Torque twisting action of the drill causes peel-up delamination, whereas drill thrust force causes push-out delamination [16]. Fig. 4.2 is a schematic representation of the two types of delamination.

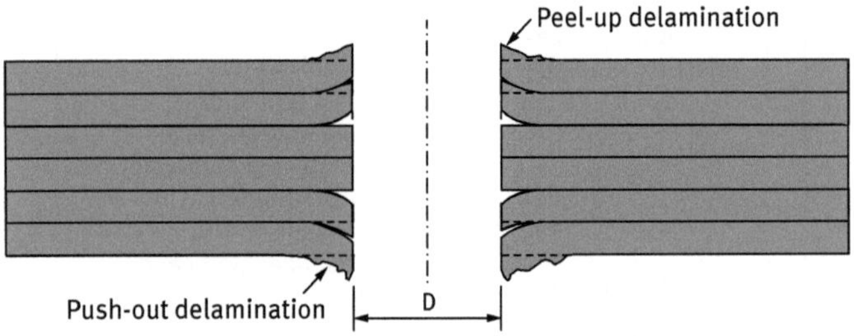

Fig. 4.2: Schematic representation of peel-up and push-out delamination.

4.2.2 Mechanisms of delamination

Drilling-induced delamination mainly occurs at the start and the finish of the drilling operation. Intermediate layers are not considered to be so affected.

4.2.2.1 Peel-up delamination

Peel-up delamination occurs at the entrance side of the hole when the cutting edge of the drill abrades the top layers of the laminate, making them move upwards, and tends to pull these abraded laminates along the flute of the tool. Thus, this material spirals up further as the drill bit progress downwards. This leads to a force acting upwards which causes separation of initial plies from the uncut laminate plies which are being pushed away by the thrust force of the drill.

4.2.2.2 Push-out delamination

Push-out delamination occurs at the exit side of the hole. The workpiece experiences compressive force during the drilling operation. Laminates under the drill tend to be drawn away from interlaminar bond around the hole. When the drilling operation nears its end, the drill is cutting the last laminate/laminates. The uncut thickness value being small, the low bond strength of this uncut laminates results in separation of plies under the applied deformation force exerted by the drill, causing the delamination. In general, the severity of push-out delamination is higher than that of peel-up delamination.

4.2.3 Methods of measuring delamination

Severity of delamination can be assessed by measuring the nature and dimensions of the delamination zone. These measurements are used to develop delamination assessment factors that facilitate the analysis of the influence of the machining parameters on the delamination of different composite materials. Major methods used in measuring delamination damage are: visual methods, image processing, acoustic emission, scanning acoustic microscopy, ultrasonic C-scan, radiography, X-ray computerized tomography, and shadow moire laser interferometer [17]. Researchers are also using machine vision systems to capture the images in high quality for further processing and measurement of the damaged zone [6].

4.2.4 Assessment of delamination

Researchers use different techniques to quantify delamination damage. Detailed discussion on these delamination factors is presented by Babu and his colleagues elsewhere [17, 18]. These factors are: Conventional delamination factor, Delamination size, Two-dimensional delamination factor (F_a), Delamination factor, Damage ratio, Adjusted delamination factor (F_{da}), Equivalent delamination factor (F_{ed}), Refined delamination factor (F_{DR}), Shape's circularity, and minimum delamination factor. The delamination factor (F_d) can also be calculated as a ratio of A_{max} to A_{nom} [19], where A_{max} is the damage area including nominal hole area (A_{nom}).

Recent works by Babu et al. [20] modified the equivalent delamination factor as the refined equivalent delamination factor (F_{red}) where an envelope enclosing the damage area is chosen to calculate the equivalent diameter. As this factor does not consider the crack length in the delamination damage assessment, Babu et al. proposed a comprehensive delamination assessment factor called effective equivalent delamination factor (FEED), which can be used for both conventional and high speed drilling [21].

Tab. 4.1: Different methods for the evaluation of delamination factor.

Sl. No.	Evaluation of delamination factor	Formula used	Scheme
1	Conventional Delamination factor (F_d)	$F_d = \frac{D_{max}}{D_0}$	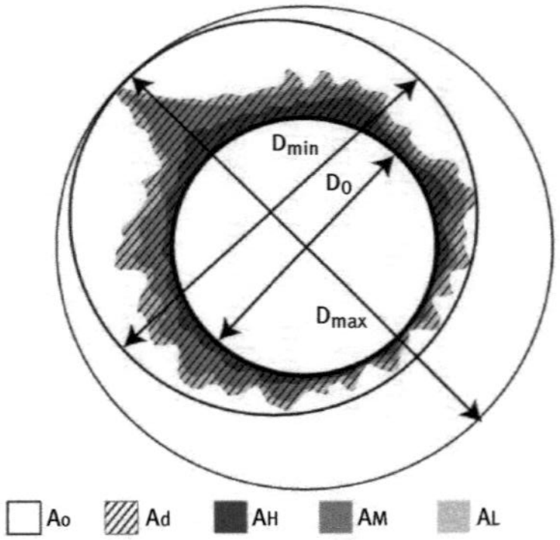
2	Delamination size	Delamination size $= R_{max} - R$	
3	Two-dimensional delamination factor (F_a)	$F_a = \left(\frac{A_d}{A_{nom}}\right)$	
4	Damage ratio	$D_{RAT} = \frac{D_{MAR}}{A_{AVG}}$	
5	Delamination factor	$F_d = \frac{A_d}{A}$	
6	Adjusted delamination factor (F_{da})	$F_{da} = F_d + \frac{A_d(F_d^2 - F_d)}{A_{max} - A_0}$	
7	Minimum delamination factor	$F_{d_{min}} = \frac{D_{min}}{D_0}$	
8	Refined delamination factor (F_{DR})	$F_{DR} = \frac{D_{max}}{D_0} + 1.783\left(\frac{A_H}{A_0}\right) + 0.7156\left(\frac{A_M}{A_0}\right)^2 + 0.03692\left(\frac{A_L}{A_0}\right)^3$	
9	Shape's circularity	$f = 4\pi\frac{A_e}{P_e^2}$	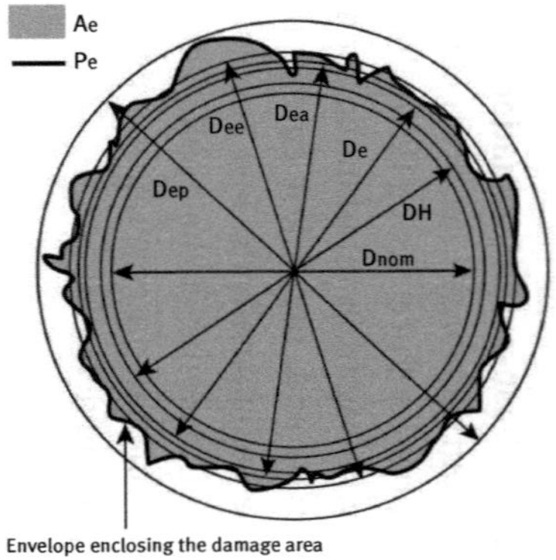
10	Equivalent delamination factor (F_{ed})	$F_{ed} = \frac{D_e}{D_0}, D_e = \left[\frac{4(A_d + A_0)}{\pi}\right]^{0.5}$	
11	Refined equivalent delamination factor	$F_{red} = \frac{D_{re}}{D_0}, D_{re} = \sqrt{\frac{4A_e}{\Pi}}$	
12	Effective equivalent delamination factor	$\text{FEED} = \frac{D_{ea} + D_{ep}}{2D_0}, D_{ea} = \sqrt{\frac{4A_e}{\Pi}}, D_{ep} = \frac{P_e}{\Pi}$	

Tab. 4.2: Different measurement methods and delamination factors used in drilling NFRCs with drilling conditions.

Delamination		Drilling conditions				Ref.
Measurement method	Assessment factor	Tool materials	D (mm)	Spindle speed (rpm)	Feed rate (mm/rev)	
Machine vision	F_d, F_{da}	HSS, twist	3,4,5	600,900,1200	0.1,0.2,0.3	[22]
Machine vision, C-scan	F_d	–	10	500, 1000, 1500, 2000	0.1, 0.2, 0.3	[13]
Enhanced radiography	Visual inspection	Brad, twist tungsten	6	28 000	0.05, 0.2	[23]
Digital image processing (Image-J software)	$F_d = (A_d + A_0)/A_0$	Twist, Jo, Parabolic	8	900, 1800, 2800	0.05, 0.12, 0.19	[19]
Machine vision, microscope	$DF = (A_d + A_0)/A_0$	Standard HSS		1000, 2000, 3000	0.1, 0.2, 0.3	[24]
Flatbed scanner, image software CorelDRAW	F_d, F_{da}	Diamond	4, 6, 8	500, 860, 1560	18, 26, 34 mm/min	[25]
Profile projector	F_d	Coated carbide spur and brad tools	6, 9, 12		0.04, 0.08	[26]
Machine vision with Rapid-I software	F_d	HSS twist drills	6, 8, 10	600, 1200, 1800	0.1, 0.2, 0.3	[27]
Tool maker's microscope	F_d	Twist 118, brad drill end mill	6	1500, 2000, 2500	0.1, 0.2, 0.3	[28]
Tool maker's microscope	F_d	Two fluted solid tungsten carbide	5	1000, 1500, 2000	0.1, 0.15, 0.2	[29]
Profile projector	F_d	HSS-Co twist drill multi-construction twist drill, brad and spur coated carbide drill	6, 9, 12	1000, 2000, 3000	0.04, 0.06, 0.08	[30]
Flatbed scanner, Image-J software	F_d	HSS twist	6	Cutting speed (mm/min): 9.42, 15.072, 20.724, 26.367	0.03, 0.06, 0.09, 0.12	[31]
Microscope	Di = Drill diameter / hole diameter	HSS	10	500, 1000, 1500, 2000	0.1, 0.2, 0.3, 0.4	[32]
Microscope	F_d	HSS twist	6, 8, 10	1000, 2000, 3000	50, 150, 250 mm/min	[33]

4.2.5 Delamination in NFRCs

Delamination due to drilling of composites and the reasons for its occurrence have been investigated extensively. But a large majority (93 %) of the studies are on synthetic fiber reinforced composites and very few on natural fiber-based composites [19]. The methods of measurement and assessment of delamination damage used in these studies along with the machining parameters are listed in Tab. 4.2.

Thrust force and cutting torque determine the incidence and extent of delamination due to drilling of composites. Their values depend on drill bit geometry, spindle speed, feed rate, drill bit material, and size [19, 34]. The traditional twist drills have been found to cause more delamination in NFRCs compared to drills like parabolic drills and brad drills which produce lower thrust forces [19, 23, 35, 36].

Another influencing factor in drilling delamination is the reinforcing fiber used in NFRCs. The nature, size, length, content percentage and pretreatment of the fibers in the composite are important in controlling their delamination [22, 31, 37–40].

Pure green composites use not only natural reinforcement fibers, the matrix too is made up from biological material, like soy protein. Even wood (example: cedar) and bamboo themselves are composite materials. Delamination due to drilling, sometimes referred to as 'fuzzing', at the entrance and exit of the drilled hole are found to occur in these as well. Thrust force of drilling directly influences the extent of damage, the maximum thrust force increasing with the feed rate [41].

It can be concluded from the investigations on drilling of NFRCs by various researchers that delamination occurs at both entry and exit, and the dominant influence for its occurrence is the feed rate during drilling. Enhancing the mechanical behavior of the composites with special treatments like alkali treatment reduces the delamination. It has also been observed that the spindle speed is the dominant factor for delamination when large drill diameters are used. Thus, the machining parameters: feed rate and cutting speed are the parameters that contribute to delamination both at the entry and exit.

4.2.5.1 Numerical prediction of delamination.

The prediction of delamination in NFRCs is based on prediction of thrust force and torque, as values of these are used for indirect assessment of delamination. Venkateshwaran and A. Perumal [13] presented a numerical prediction of delamination using regression equations.

Multivariable regression analysis is used to derive the relationship between the delamination factor (F_d) and the cutting parameters (speed and feed) in the drilling of the banana/epoxy composite.

The relationship between the delamination factor and the cutting parameters are given by equations (4.1) and (4.2).

$$F_{d,\text{entry}} = 1.021 + 0.000479S + 0.0886F \quad (4.1)$$

$$F_{d,\text{exit}} = 1.029 + 0.000167S + 0.0606F \quad (4.2)$$

Comparison of the experimental delamination factor (F_d) with the predicted delamination factor shows that the maximum error percentages between the experimental delamination factor and the regression model at entry and exit are 0.678 and 0.357, respectively. Very few studies with FEM simulations of drilling of NFRCs are available.

4.2.6 Geometrical errors

Surface roughness

Performance of a machined hole is significantly influenced by the surface finish obtained in drilling. This facilitates better integrity with the fasteners in assembly. Research has been carried out to optimize the process parameters in order to achieve a desired surface finish [42]. It has been found that surface roughness (R_Z) increases with increase in feed rate and spindle speed. At high spindle speeds, R_Z values remain constant even at high feed rates [43].

Experiments conducted using various drill geometries such as high speed steel (HSS) twist drill, multiconstruction twist drill, and brad and spur types on hybrid composites using different natural fibers: glass, sisal, and jute fibers as reinforcing material show that the brad and spur drill produce less induced damage around the hole surface and generate less thrust force than the other two drills. The researchers examined the hole at the interfaces between the layers by SEM. Fiber breakage and fiber pull-out are observed at the drilled surface [23, 30].

Hole size error

Accuracy of the drilled hole has a significant effect on the performance and facilitates better assembly. Alkali-treated fibers in NFRCs are found to show better dimensional accuracy than the composite specimen with other fiber content and treatments [39]. Fig. 4.3 illustrates the variation of hole diameter with different fiber content. The composite specimen with 30 wt% fiber content and 8 h treatment shows better dimensional accuracy than composites with other fiber content and treatment [39].

4.2.7 Thermal damages

These defects are due to the high temperature generated during drilling. Moderate temperatures generated during drilling may be beneficial in softening the matrix and

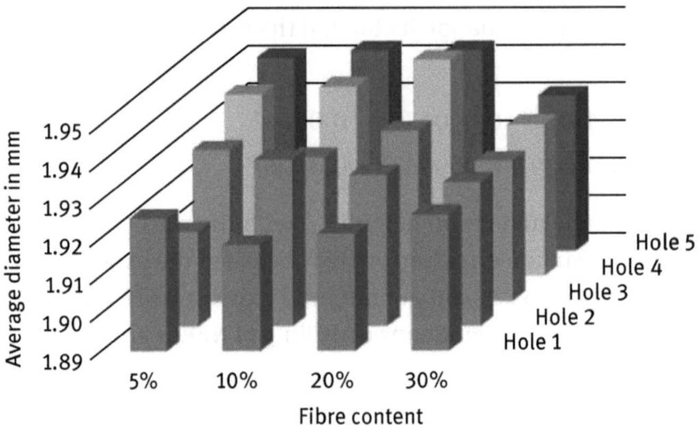

Fig. 4.3: Drill hole profile analysis of hybrid composite with different amounts of fiber content and 8 h treatment.

thereby reducing thrust force. This results in a decrease in delamination. But higher temperatures nearer or above the glass transition temperature reduce the strength of the composite material and may cause an increase in delamination [16]. The thermomechanical stress developed during drilling by the generation of temperature and induced forces cause elongation or movement of polymer chains resulting in a large damaged zone [19]. A comprehensive analysis of the heat generation and distribution in the tool-workpiece interaction zone is necessary to completely understand the drilling-induced damage in the thermoplastic composite laminates, both qualitatively and quantitatively. Similar opinion was raised by Khasaba that the form in which the thermal and mechanical damages occur, and the correlations between them have not yet been fully reported and problems remain to be solved [44].

4.3 Machining defects in milling

Milling is one of the machining/finishing operations for removing extra material from the workpiece to obtain desired surface quality with dimensional tolerances. As components made with composite materials are generally made to near net shapes, milling is a rarely used machining process in composites [9]. End-milling is the preferred machining process for producing precision pockets and slots in various materials. Milling operations on NFRCs also induce damage in the materials such as microcracks, matrix burning and delamination. These defects affect the performance of the components.

4.3.1 Delamination in milling

Delamination is treated as a severe defect in milling; its severity depends on the cutting parameters used during the milling process. Delamination in milling is assessed by a delamination factor [8] which is defined as the ratio of the width of maximum damage, (W_{max}), and the nominal width of the cut (W) as shown schematically in Fig. 4.4. Equation (4.3) represents the formula for calculating the delamination factor.

$$F_d = \frac{W_{max}}{W}. \tag{4.3}$$

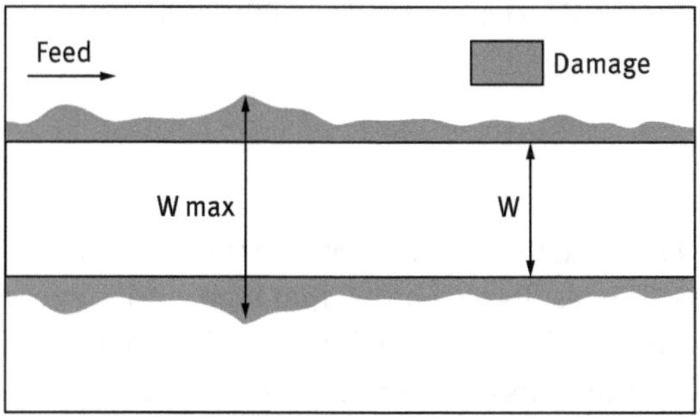

Fig. 4.4: Schematic representation of delamination factor in milling.

Research reports [8] indicate that delamination damage is less in natural fiber reinforced composites when compared with synthetic fiber reinforced composites. The delamination factor decreases with the cutting speed and increases with the feed rate. The results also reveal that hemp fiber reinforced composites give better performance than the other fiber reinforced composites investigated and jute fiber reinforced composites give larger values of delamination factor (F_d). This is due to the differences in the geometries and physical properties of different types of natural fibers.

4.3.1.1 Numerical prediction of delamination in milling

Researchers [8] used multiple linear regressions using Minitab 16 software to derive the relationship between the delamination Factor (F_d) and the cutting parameters (speed and feed) in milling different natural fibers (hemp, jute, banana) and synthetic fiber (glass).

The relationship between the delamination factor and the cutting parameters are given by equations (4.4) to (4.7).

$$F_{d,GFRP} = 1.20 - 0.00434S + 0.671F \tag{4.4}$$

$$F_{d,HFRP} = 1.15 - 0.00388S + 0.448F \tag{4.5}$$

$$F_{d,JFRP} = 1.22 - 0.00756S + 0.958F \tag{4.6}$$

$$F_{d,BFRP} = 1.23 - 0.00705S + 0.782F \tag{4.7}$$

Obtained adjusted R values for equations (4.4), (4.5), (4.6), and (4.7) are 90.6 %, 89.5 % 91.4 %, and 96.6 % and indicate the accuracy of prediction. Comparison of the experimental delamination factor (F_d) with the predicted delamination factor shows that the maximum error percentage between the experimental delamination factor and the regression model is less than 5 %.

4.3.2 Geometrical errors

Surface finish and tribological properties are important for the performance of the milled product. Several researchers [8, 45, 46] focus on the effect of the milling process on tribological aspects.

Surface roughness

Research work indicates that the feed rate and the cutting speed are the dominant factors that influence surface roughness. The results also reveal that high cutting speed and low feed rate result in lower surface roughness for milling kenaf fiber reinforced composites. The machined surface of natural fiber reinforced plastics (NFRPs) is significantly dependent on the fiber stiffness and interface bonding. These results match those from experiments on composites with natural fibers (hemp, jute, banana) and synthetic fiber (glass).

In experiments [46] to study the effect of milling on surface tribology, the roughness values before and after the profile milling process are compared. In these, a term called global roughness gain ratio is used, which is the ratio of difference of R_a values before and after milling to the R_a value before milling, and it is expressed as a percentage. The results reveal that the final roughness value can be up to four times the initial roughness value in the case of sisal fibers. Bamboo and miscanthus also show the same behavior in terms of induced machining roughness. The feed rate is found not to have a significant effect on the global surface roughness gain ratio. The

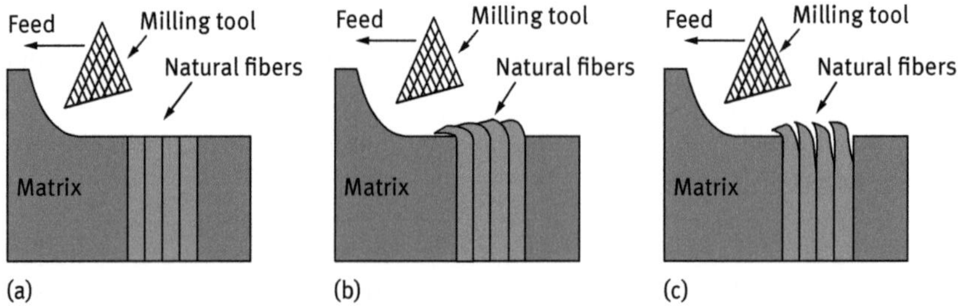

Fig. 4.5: Principal mechanisms when cutting NFRPs; (a) ideal shearing of fibers, (b) real shearing of fibers, (c) real shearing of fibers showing interfaces break [46].

observations with SEM and energy analysis reveal that sisal fibers, which have the lowest cutting contact stiffness, contribute more to the increase in surface roughness after machining because the fiber extremities that exceed the machining surface still leaned in the feed direction (Fig. 4.5 (b)). In addition, the low fiber-fiber adhesion can lead to the breakage of the interfacial liaison between elementary fibers during milling as observed on sisal NFRP (Fig. 4.5 (c)).

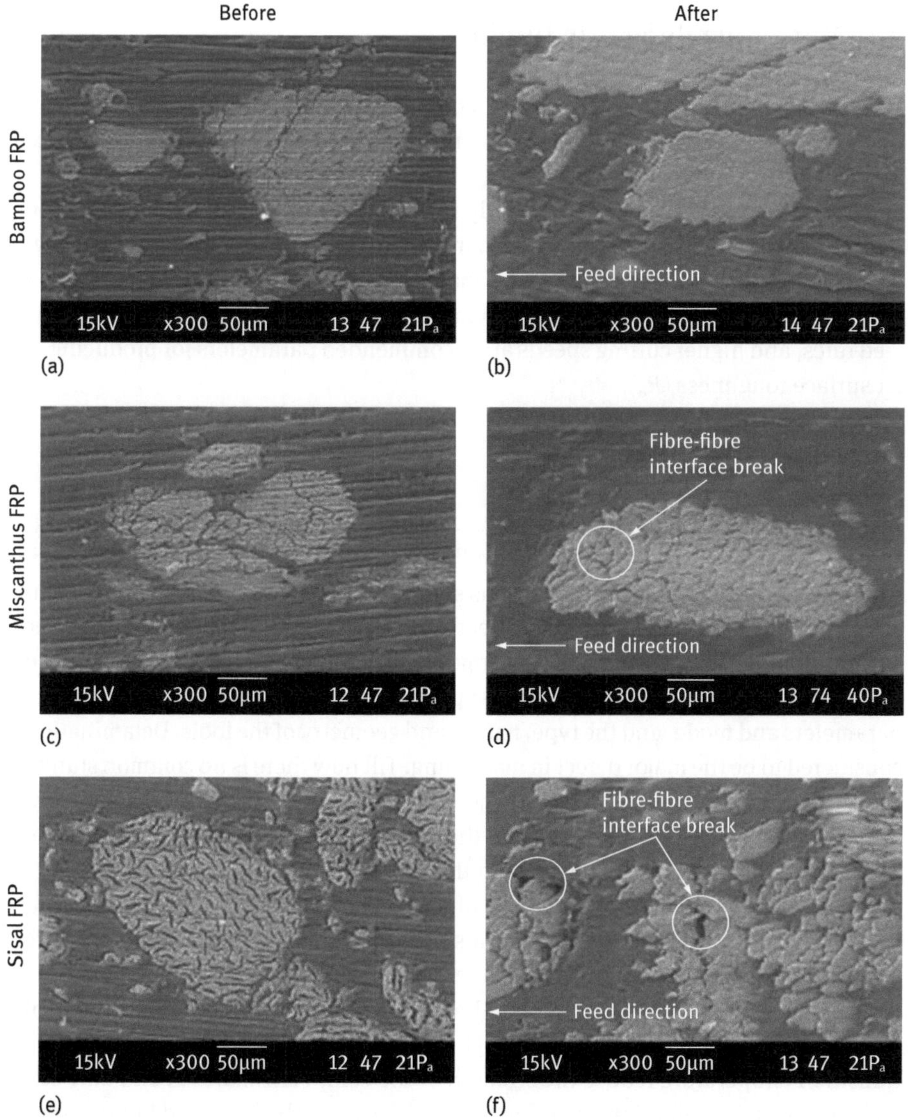

Fig. 4.6: SEM images of surface state of NFRP before and after profile milling process with feed rate 0.04 mm/tooth [46].

Fig. 4.6 compares the SEM of surface states of workpieces before and after profile milling operations. The initial state before machining shows polishing streaks and fiber debris and fiber sections appear as composed of the groups of several elementary fibers. Different morphologies can be observed for different NFRCs due to the activated mechanisms during the profile milling operation. Different composites exhibit different severity of fiber interface failures leading to different morphologies.

4.4 Machining defects in turning

Turning is generally not needed as NFRC components are produced by net shape manufacturing. However, turning operation is needed for finishing and dimensional accuracy.

Experiments [47] conducted to study the influence of cutting parameters on the surface roughness obtained in turning parts made from HDPE matrix reinforced with wood fibers with HSS cutting tool reveal that surface roughness (R_a) increases with increasing feed rate. Further research [48] indicates that higher tool nose radius, lower feed rates, and higher cutting speeds are recommended parameters for producing better surface roughness (R_a).

4.5 Summary

Natural fibers, due to their abundant availability and biodegradability, are replacing expensive, nonrenewable and hazardous synthetic fibers in composites. The machinability of natural fiber composites is directly related to the characteristics of the fibers such as their mechanical, physical, thermal, and chemical properties. Machining defects of NFRCs depend on several factors: primary manufacturing process, machining parameters and mode, and the type, nature and geometry of the tools. Delamination is considered to be the major defect in machining. Till now there is no common standard in assessing the delamination factor. There are several techniques or methods developed to measure the delamination, and the delamination factor obtained by different techniques may differ. Use of higher spindle speed, lower feed rate, smaller drill bit diameter, parabolic drill, and point angle of 118° can reduce the delamination. Surface roughness and hole size accuracy can also be adversely affected. Temperature generated during machining plays a vital role. Moderate temperatures generated during drilling may be beneficial in softening the matrix thereby reducing thrust force and, in turn, reducing delamination. But higher temperatures nearer or above the glass transition temperature reduce the strength of the composite material and may cause an increase in delamination and other defects like matrix burning and hole size error. Even though drilling is the most common process for machining NFRCs, milling and turning are also carried out to meet accuracy and dimensional requirements.

References

[1] Soutis C. Fibre reinforced composite in aircraft construction. Progr Aerosp Sci. 2005;41:143–151.

[2] Nagarajan VA, Selwin Rajadurai J, Annil Kumar T. A digital image analysis to evaluate delamination factor for wind turbine composite laminate blade. Compos Part B. 2012;43:3153–3159.

[3] Lopez de Lacalle LN, Lamikiz A, Campa FJ, et al. Design and test of multitooth tool for CFRP milling. J Compos Mater. 2009;43:1791–1799.

[4] Ying-Chen Z, Hong-Yan W, Yi-Ping Q. Morphology and properties of hybrid composites based on polypropylene/polylactic acid blend and bamboo fiber. Bioresource Technol. 2010;101:7944–7950.

[5] Singh I, Bajpaj PK, Malik D, et al. Feasibility study on microwave joining of green composites. Akademeia. 2011;1:1–6.

[6] Ramesh M, Palanikumar K, Reddy KH. Mechanical property evaluation of sisal-jute-glass fiber reinforced polyester composites. Compos B. 2013;48:1–9.

[7] Nassar MMA, Arunachalam R, Alzebdeh KI. Machinability of natural fiber reinforced composites: a review, Int J Adv Manuf Technol. 2017;88:2985–3004.

[8] Babu GD, Babu KS, Gowd BU. Effect of machining parameters on milled natural fiber-reinforced plastic composites. J Adv Mech Eng. 2013;1:1–12.

[9] Davim PJ, Silva LR, Festas A, Abrão AM. Machinability study on precision turning of PA66 polyamide with and without glass fiber reinforcing. Mater Des. 2009;30:228–234.

[10] Abrao AM, Faria PE, Campos Rubio JC, Reis P, Davim JP. Drilling of fiber reinforced plastics: a review. J Mater Process Technol. 2007;186:1–7.

[11] Salleh Z, Berhan MN, Hyie KM, Taib YM, Kalam A, Roselina NRN. Open hole tensile properties of kenaf composite and kenaf/fibreglass hybrid composite laminates. Proc Eng. 2013;68:399–404.

[12] Lachaud F, Piquet R, Collembet F, Surchin L. Drilling of composite structures. J Compos Struct. 2001;52(3–4):511–516.

[13] Venkateshwaran N, ElayaPerumal A. Hole quality evaluation of natural fiber composite using image analysis technique. J Reinf Plast Compos. 2013;32:1188–1197.

[14] Wong TL, Wu SM, Croy CM. An analysis of delamination in drilling composite materials. In: Proceedings of 14th SAMPE Technology Conference, Atlanta, GA, USA; 1982. p. 471–483.

[15] Tsao CC, Hocheng H. Evaluation of thrust force and surface roughness in drilling composite material using Taguchi analysis and neural network. J Mater Process Technol. 2008;203:342–348.

[16] Kennametal (2011). Composite machining. https//wwww.kennametal.com. Composite machining guide for Aerospace pdf A34–55.

[17] Babu J, Philip J, Zacharia T, Davim JP. Delamination in composite materials: measurement, assessment and prediction. In: Davim JP. Machinability of fibre-reinforced plastics. Berlin: De Gruyter; 2015. p. 139–162.

[18] Babu J, Sunny T, Paul NA, Mohan KP, Philip J, Davim JP. Assessment of delamination in composite materials: a review. Proc Inst Mech Eng B J Eng Manuf. 2016;230(11):1990–2003.

[19] Bajpai PK, Debnath K, Singh I. Hole making in natural fiber-reinforced polylactic acid laminates: an experimental investigation. J Thermoplast Compos Mater. 2017;30(1):30–46.

[20] Babu J, Paul NA, Mohan KP, Philip J, Davim JP. Examination and modification of equivalent delamination factor for assessment of high speed drilling. J Mech Sci Tech. 2016;30(11):5159–5165.

[21] Babu J, Alex PN, Abraham SP, Philip J, Anoop BN, Davim JP. Development of comprehensive delamination assessment factor and its evaluation with high-speed drilling of composite laminates using a twist drill. Proc IMechE Part B: J Eng Manuf. 2017:1–13.

[22] Chandramohan D, Marimuthu K. Drilling of natural fiber particle reinforced polymer composite material. Int J Adv Eng Res Stud. 2011;1:134–145.
[23] Durão LMP, Gonçalves DJS, Tavares JMRS, de Albuquerque VHC, Panzera TH, Silva LJ, Vieira AA, Baptista APM. Drilling delamination outcomes on glass and sisal reinforced plastics. Mater Sci Forum. 2013;730–732:301–306.
[24] Patel K, Gohil PP, Chaudhary V, Patel K. Investigation on drilling of banana fibre reinforced composites. International Conference on Civil, Materials and Environmental Sciences (CMES 2015). 1:201–205.
[25] Abilash N, Sivapragash M. Optimizing the delamination failure in bamboo fiber reinforced polyester composite. J King Saud Univ Eng Sci. 2016;28:92–102.
[26] Ramesh M, Palanikumar K, Reddy KH. Influence of tool materials on thrust force and delamination in drilling sisal-glass fiber reinforced polymer (S-GFRP) composites. Proc Mater Sci. 2014;5:1915–1921.
[27] Balaji NS, Jayabal S, Kalyana Sundaram S, Rajamuneeswaran S, Suresh P. Delamination analysis in drilling of coir-polyester composites using design of experiments. Adv Mater Res. 2014;984–985:185–193.
[28] Azuan SAS, Juraidi JM, Muhamad WMW. Evaluation of delamination in drilling rice husk reinforced polyester composites. Appl Mech Mater. 2012;232:106–110.
[29] Babu GD, Babu KS, Gowd BU. Optimization of machining parameters in drilling hemp fiber reinforced composites to maximize the tensile strength using design experiments. Indian J Eng Mater Sci. 2013;20:385–390.
[30] Ramesh M, Palanikumar K, Reddy KH. Experimental investigation and analysis of machining characteristics in drilling hybrid glass-sisal-jute fiber reinforced polymer composites. 5th International & 26th All India Manufacturing Technology, Design and Research Conference (AIMTDR 2014), December 2–7.
[31] Sridharan V, Muthukrishnan N. Optimization of machinability of polyester/modified jute fabric composite using grey relational analysis (GRA). Proc Eng. 2013;64:1003–1012.
[32] Vinayagamoorthy R, Rajeswari N, Vijayshankar S, Vivekanandan M, Bellala SRM, Subramaniam KRV. Surface and subsurface analysis of hybrid polymer composites during machining operations. Proc Mater Sci. 2014;5:2075–2083.
[33] Aravindh S, Umanath K. Delamination in drilling of natural fibre reinforced polymer composites produced by compression moulding. Appl Mech Mater. 2015;766–767:796–800.
[34] Yallew TB, Kumar P, Singh I. A study about hole making in woven jute fabric-reinforced polymer composites. Proc Inst Mech Eng L: J Mater: Des Appl. 2015;0:1–11. doi: 10.1177/1464420715587750
[35] Bajpai PK, Singh I. Drilling behavior of sisal fiber-reinforced polypropylene composite laminates. J Reinf Plast Compos. 2013;32:1569–1576.
[36] Jayabal S, Natarajan U, Sekar U. Regression modeling and optimization of machinability behavior of glass-coir-polyester hybrid composite using factorial design methodology. Int J Adv Manuf Technol. 2011;55:263–273.
[37] Jayabal S, Velumani S, Navaneethakrishnan P, Palanikumar K. Mechanical and machinability behaviors of woven coir fiber-reinforced polyester composite. Fibers Polym. 2013;14:1505–1514.
[38] Balaji NS, Jayabal S, Kalyana Sundaram S, Rajamuneeswaran S, Suresh P. Delamination analysis in drilling of coir-polyester composites using design of experiments. Adv Mater Res. 2014;984–985:185–193.
[39] Athijayamani A, Thiruchitrambalam M, Natarajan U, Pazhanivel B. Influence of alkali-treated fibers on the mechanical properties and sisal fiber hybrid polymer composite. Polym Compos. 2010;723–731.

[40] Babu D, Babu KS, Gowd BUM. Drilling uni-directional fiber-reinforced plastics manufactured by hand lay-up influence of fibers. Am J Mater Sci Technol. 2012;1:1–10.
[41] Mizobuchi A, Takagi H, Sato T, Hino J. Drilling machinability of resin-less “green” composites reinforced by bamboo fiber. WIT Transactions on the Built Environment. 2008;9:185–194.
[42] Davim JP, Mata F. Optimisation of surface roughness on turning fibre-reinforced plastics (FRPs) with diamond cutting tools. Int J Adv Manuf Technol. 2005;26:319–323.
[43] Hutyrová Z, Zajac J, Michalik P, Mita D, Dupľák J, Gajdoš S. Study of surface roughness of machined polymer composite material. Int J Poly Sci. 2015; Article ID 303517, 6 pages. doi: 10.1155/2015/303517.
[44] Khashaba UA. Drilling of polymer matrix composites: A review. J of Compos Maters. 2012;47(15):1817–1832.
[45] Harun A, Haron HC, Ghani JA, Mokhtar S, Ting ST. Study the effect of milling parameters on surface roughness during milling kenaf fibre reinforced plastic. Adv Environ Biol. 2015;9:46–52.
[46] Chegdani F, Mezghani S, El Mansori M, Mkaddem A. Fiber type effect on tribological behavior when cutting natural fiber reinforced plastics. Wear. 2015;332:772–779.
[47] Somsakova Z, Zajac J, Michalik P, Kasina M. Machining of wood plastic composite (pilot experiment). Mater Plast. 2012;49(1):55–57.
[48] Zajac J, Hutyrová Z, Orlovský I. Investigation of surface roughness after turning of one kind of the bio-material with thermoplastic matrix and natural fibers. Adv Mater Res. 2014;941–944:275–279.

M. A. A. Aldahdooh, Ali Alnuaimi, A. Jamrah, N. Muhamad Bunnori,
M. A. Megat Johari, and M. I. Martini

5 GUSMRC – From concept to structural application

5.1 Introduction

Overloading is considered one of the most predominant degradation mechanisms of reinforced concrete (RC) structures, which could lead to serious structural failure and collapse in extreme cases [1, 2]. Retrofitting material that has good compatibility with the parent material of the damaged concrete structure is needed [3].

Among the recent and promising retrofitting techniques, the externally bonded, ultra-high performance, fiber reinforced, cementitious composite (UHPFRCC) strip techniques (e.g., CARDIFRC) are used for improving the structural behavior of damaged concrete structures [3]. However, its major drawback is the utilization of high cement content and its mix design process that to date is still unstandardized. Increasing the cement content beyond the limit values will not enhance the properties of concrete [4, 5], but it will increase the emission of greenhouse gases that affect global warming [6]. Therefore, GUSMRC has been developed to overcome some of these problems (Fig. 5.1).

GUSMRC is a new class of green, ultra-high performance, fiber reinforced, cementitious composites (GUHPFRCCs) in which 50 % of the volume contains ultrafine palm oil fuel ash (UPOFA). It has been developed at the Universiti Sains Malaysia (USM) by Aldahdooh et al. [7]. GUSMRC has been designed using two ideal and practical methods: (1) by optimizing the mix design of traditional UHPFRCCs using response surface methodology (RSM) and absolute volume (AVM) methods [8]; (2) by replacing a large portion of the binder in the UHPFRCC optimum mix with UPOFA [7, 9] as shown in Fig. 5.1. Due to its excellent properties, GUSMRC has been applied as a green retrofitting material for improving the performance of existing concrete structures damaged by overloading [10].

5.2 UHPFRCCs: Definition and constituent materials

UHPFRCC is an advanced reinforced cementitious material and is one of the UHPFRC types [11, 12]. In terms of mechanical performance, UHPFRCC is characterized by an excellent compressive strength, tensile strength, bending tensile strength, elastic modulus, energy absorption capacity, and elastic post-cracking bending strength. In terms of durability, UHPFRCC shows an extremely dense microstructure (negligible water adsorption, water and gas permeability, and porosity) and an extremely low diffusion

https://doi.org/10.1515/9783110435788-006

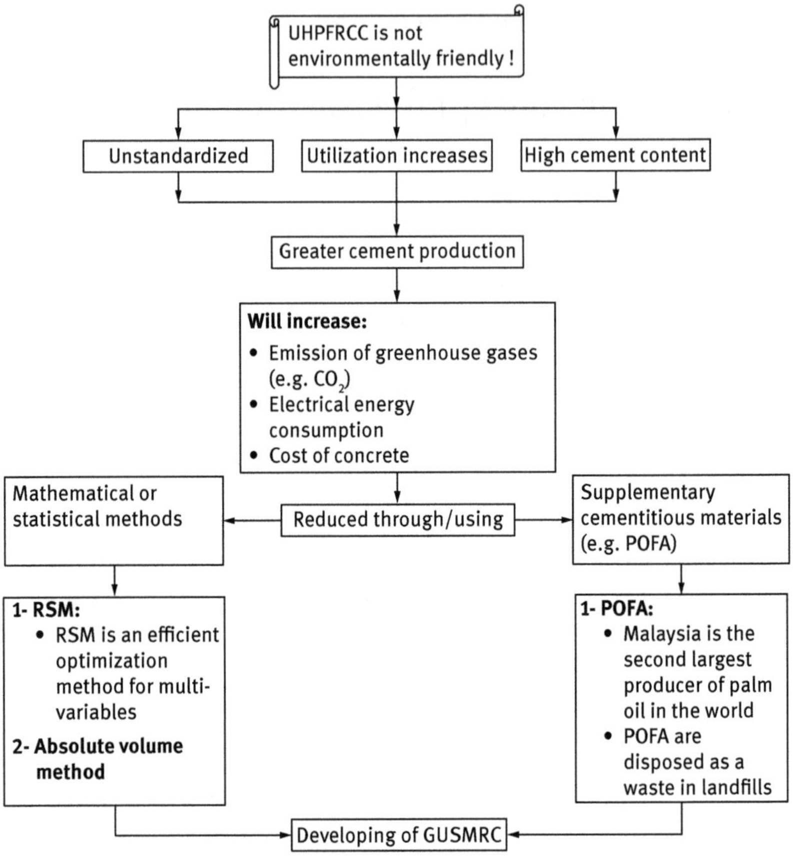

Fig. 5.1: Reasons for developing GUSMRC [7–10].

coefficient [11, 13–17] as tabulated in Tab. 5.1. Meanwhile, in terms of sustainability, this type of concrete still needs to be evaluated with regard to its high binder content (especially the cement content) relative to the regularly used mixtures [17].

Generally, UHPFRCC is characterized as a composite that has a high content of cement and silica fume, large volume of steel fiber, high dosage of high-range water-reducing admixture, low water/binder ratio, and the absence of coarse aggregates that are larger than 4 mm [18]. Tab. 5.1 shows a summary of the ranges of UHPFRCC compositions and average material properties. One example of UHPFRCC is known under the trade name CARDIFRC [3, 19, 20].

Tab. 5.1: The range of UHPFRCC compositions and average mechanical properties.

		UHPFRCCs (kg/m³)	CARDIFRC [3, 13, 19, 20]
Matrix composition	Portland cement	(700–1000) [11, 21]	744–855
	Coarse aggregate	(0–200) [11, 21]	0
	Fine aggregate	(1000–2000) [11, 21]	940–1173
	Silica fume	(200–300) [11, 21]	178–214
	Water	(110–200) [11, 21]	149–188
	Superplasticizer	(9–71) [22]	28–55
	Reinforcement/fibers	(> 150) [11, 21]	468
	Water/cement ratio	(< 0.24) [11, 21]	0.20–0.22
	Water/binder ratio	(< 0.22) [11, 21]	0.16–0.18
	Superplasticizer/cement ratio	(0.018–0.051) [22]	0.033–0.074
	Silica fume/cement ratio	(> 0.25) [11, 21]	0.24–0.25
Properties	Compressive strength (MPa)	(> 150) [11, 21]	> 185
	Tensile strength (MPa)	(> 7) [23]	12–13.5
	Modulus of elasticity (GPa)	(50–70) [11, 21]	> 48
	Splitting tensile strength (MPa)	(> 18) [24]	24–25
	Flexural strength (MPa)	(> 25) [11, 14]	> 30

5.3 POFA: definition and its applications in concrete production

POFA is a by-product of burning, obtained by burning of fibers, shells and empty fruit bunches as fuel in palm oil mill boilers [25, 26]. The palm oil industry is one of the major agro-industries in countries such as Malaysia, Indonesia, and Thailand. Moreover, high amounts of POFA are produced in Malaysia [27]. Malaysia is the second largest producer of palm oil after Indonesia. Furthermore, palm oil production in Malaysia has increased over the years; it reached 16.9 million tons in 2010 and 18.9 million tons in 2011 [28, 29].

Most POFA is disposed of as waste in landfills, which may contribute to environmental problems in the future [30]. Therefore, a lot of research works have been conducted to find a suitable solution for proper POFA disposal.

Many researchers have found that POFA has pozzolanic qualities and properties in concrete. In fact, POFA can be considered as pozzolanic material [31–33]. In the early 1990s, Tay utilized POFA as partial cement replacement. He indicated that POFA has low pozzolanic properties. Moreover, Tay [34] recommended that POFA is not suitable for use as a partial cement replacement in a quantity that exceeds 10 % of cement mass. However, Awal and Hussin [27] showed that POFA can be utilized as a supplementary cementitious material, and POFA has a high potential in suppressing expansion associated with alkali–silica reaction in concrete.

POFA has been utilized in many fields, especially in civil and environmental fields. POFA has been observed to enhance the transport properties of concrete. POFA reduces heat development [30] and improves concrete resistance to chloride ion penetration [33–35], acidic environment [34], and sulfate attack [26]. Moreover, POFA also enhances engineering properties, such as the flexural performance of engineered cementitious composites (ECC) [36], thereby improving normal concrete compressive performance [37, 38]. Furthermore, POFA has been utilized in high performance concrete (HPC) production, the highest compressive strength was found in the range 60–86 MPa, which was obtained at POFA (with median particle size of approximately 10 μm) replacement level of 20 % at day 28 with 550 kg/m^3 to 560 kg/m^3 total binder [32, 39, 40].

Megat Johari et al. [33] modified the treatment and grinding process of POFA by heat treatment to remove the excess carbon content and to decrease the POFA median particle size to approximately 2.06 μm. A highly efficient pozzolan was obtained through their treatment processes as shown in Fig. 5.2. The modified ultrafine POFA (UPOFA) was utilized for improving the engineering and transport properties of high-strength green concrete (HSGC). They found that the compressive strength could exceed 95 MPa after 28 days with a replacement level of up to 60 % for HSGC production that contains high UPOFA content. Furthermore, the use of UPOFA (particularly in high volume) can contribute to a healthier and more sustainable environment, which increases green concrete products and may reduce concrete cost.

Recently, Aldahdooh et al. [7] reduced the binder content by replacing greater portions of the cement and silica fume in UHPFRCCs with UPOFA as supplementary cementitious materials while generally maintaining its mechanical properties. The result was the production of a GUHPFRCC (**GUSMRC**) with low cement content of 360.25 kg/m^3 and with ultra-high compressive strength of 158.28 MPa.

5.3.1 UPOFA treatment procedure

The treatment process is summarized in five steps as shown in Fig. 5.2. Firstly, the original POFA was dried before the sieving process at 100 ± 5 °C for 24 h using an electric oven to facilitate the sieving process and then sieved by passing a 300 μm sieve to remove impurities, such as kernel shells, and unburned fibers, after removing moisture content during the drying step. After that, the sieved POFA was ground using a ball mill to increase the surface area and to reduce the particle size to improve its poz-

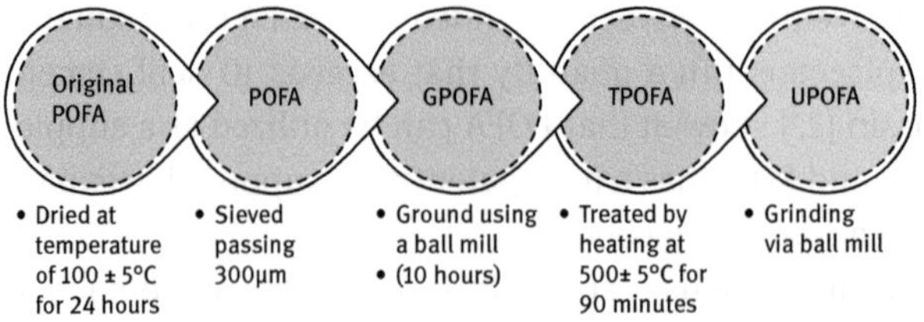

Fig. 5.2: UPOFA preparation procedures according to Megat Johari et al. [33].

zolanic properties and reactivity and then it was heat treated by heating at 500 ± 5 °C for 90 min to eliminate unburned carbon and to avoid glassy phase crystallization and particle agglomeration. Finally, the heat-treated POFA was subjected to further grinding by the same laboratory ball mill to obtain the ultrafine-POFA (**UPOFA**).

5.4 GUSMRC definition, development, and its applications

As mentioned earlier, GUSMRC is a new class of green UHPFRCCs, in which 50 % of the volume contains UPOFA [7–9]. GUSMRC is characterized by high compressive strength (up to 158 MPa), flexural strength (in excess of 46.69 MPa), direct tensile strength (up to 13.78 MPa), splitting strength (in excess of 20 MPa), high waste materials content (up to 290 kg/m^3) and low cement content (less than 360 kg/m^3) [7, 9] as tabulated in Tab. 5.3. Due to these excellent properties, it has been applied as a green retrofitting material for improving the performance of existing concrete structures damaged by overloading [10].

The key advantage of the GUSMRC mix is that, unlike CARDIFRC mixes, its cement content is less than 360 kg/m^3, and this may lead to a solution for alleviating the environmental and economic effects of cement and POFA [10].

5.4.1 Development process

Fig. 5.3 describes the development procedure of GUSMRC, which comprised the following stages:

Stage I: Optimizing the binder content in the traditional UHPFRCC mixture to yield acceptable mechanical strength using RSM and AVM methods [8].

Stage II: Using the optimum traditional UHPFRCC mix as control mix for GUSMRC development by partial binder replacement with UPOFA as shown in Fig. 5.3 [7, 9].

Stage III: Applying GUSMRC as a green retrofitting material for improving the structural behavior of damaged RC beams [10].

5.4.1.1 Optimum mix constituents and properties of traditional UHPFRCC

A practical method was developed for adjusting UHPFRCC binder content using RSM [41] and AVM methods according to ACI-211.1 [45] guidelines as presented in Fig. 5.3.

An optimal predictor quadratic model (equation (5.1)) was used in determining the optimal condition for responses using RSM [41–44].

$$Y = \beta_0 + \sum_{i=1}^{k} \beta_i X_i + \sum_{i=1}^{k} \beta_{ii} X_i^2 + \sum_{i_{i \le j}}^{k} \sum_{j}^{k} \beta_{ij} X_i X_j + e_i, \tag{5.1}$$

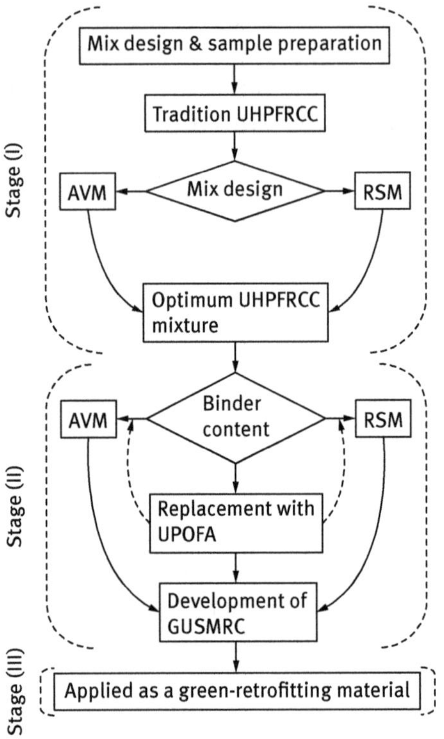

Fig. 5.3: Development stages of GUSMRC.

where Y is the predicted response, X_i and X_j are the coded values of preparation variables, i is the linear coefficient, j is the quadratic coefficient, β is the regression coefficient, k is the number of factors studied and optimized in the experiment, and e is the random error.

Considering the effect of cement (OPC) and microsilica (DSF) contents on mechanical strength, RSM was applied to adjust the amount of OPC and DSF components based on reasonable ranges. The remaining components (water, steel fibers and superplasticizer) were determined from preliminary study and the available literature (Tab. 5.1) as given in Tab. 5.2. In the final stage, the following equation was used to determine the fine aggregate content according to ACI-211.1 [45] guidelines:

$$\text{Sand}\left(\frac{\text{kg}}{\text{m}^3}\right) = \left[1 - \left(\sum_{k=1}^{n} (V)_k\right)\right] \rho_s, \tag{5.2}$$

where V_k is the volume fraction of each component k, $(\rho)_s$ is the sand density, and n is the number of mixture components excluding the sand. The results of the optimum traditional UHPFRCC mix are given in Tab. 5.2.

Tab. 5.2: Optimum traditional UHPFRCC mix constituents and properties [8].

Components			(kg/m³)
Cement			720.49
Microsilica			214.25
Mining sand			1057.31
Water			168.30
Superplasticizer			50.43
Steel fibers	l_1 = 6 mm		390
	l_2 = 13 mm		78
Water/cement			0.23
Water/binder			0.18
Microsilica/cement			0.30
Mechanical properties (28 days)		Compressive strength (MPa)	181.41
		Direct tensile strength (MPa)	12.49
		Flexural strength (MPa)	30.31
		Splitting tensile strength (MPa)	18.04
		Modulus of elasticity (GPa)	49.37

5.4.1.2 Optimum mix constituents and properties of GUSMRC

The optimum traditional UHPFRCC mix (Tab. 5.2) was used as control mix for GUSMRC development as presented in Tab. 5.3. UPOFA was utilized as partial substitute for the OPC and DSF of traditional UHPFRCC on a volume-volume basis according to ACI-211.1 [45] guidelines and using RSM. Then the binder contents (kg/m³) of POC, DSF and UPOFA for the optimum mix were determined using the Absolute Volume Method (AVM), which depends on the densities of the materials. The other ingredients of the control traditional UHPFRCC mix remained the same to ensure that any changes in the engineering properties of the GUSMRC were due to the partial replacement levels of OPC and DSF by UPOFA.

5.4.2 Tensile behavior of GUSMRC

The tensile behavior of GUSMRC **was** investigated as shown in Fig. 5.4. The behavior of GUSMRC under tensile strength can be described in three phases, which are linear elastic phase, strain hardening phase, and strain softening phase. Tab. 5.4 summarized the crack behavior at each phase of strain.

Tab. 5.3: Optimum GUSMRC mix constituents and properties [7, 9].

Components		**(kg/m³)**
Cement		360.25
Microsilica		214.25
UPOFA		290.52
Mining sand		1057.3
Water		168.30
Superplasticizer		50.43
Steel fibers	$l_1 = 6$ mm	390
	$l_2 = 13$ mm	78
Water/binder		0.195
Microsilica/binder		0.247
Mechanical properties (28 days)	Compressive strength (MPa)	156.72
	Direct tensile strength (MPa)	13.35
	Flexural strength (MPa)	42.38
	Splitting tensile strength (MPa)	20.46
	Modulus of elasticity (GPa)	46.72

Tab. 5.4: Tensile behavior of GUSMRC [9].

Phase	**Strain (ε)**	**Crack mode**
Phase (I)	0–200 μ strain	Microcracks
Phase (II)	200–400 μ strain	
Phase (III)	> 400 μ strain	Macrocracks

5.4.3 GUSMRC application

The optimum GUSMRC mixture was applied as a green retrofitting material to enhance the mechanical behavior of overloading-damaged reinforced concrete (RC) beams. The GUSMRC was cast in strip molds (1030 mm long and 100 mm wide molds with two different thicknesses [16 and 20 mm]) according to Alaee and Karihaloo [3] and was used as externally bonded strips as shown in Fig. 5.5. CONCRESIVE® 1441S epoxy bonding paste was used in this study to bond the GUSMRC retrofitted strips with the prepared surfaces of the damaged beams a shown in Fig. 5.5 (a) and (b).

Results show that GUSMRC can be used as a new green retrofitting material for the repair, rehabilitation, and upgrade of damaged RC beams. GUSMRC not only increases the load carrying capacity over that of the control beams, but also significantly improves the serviceability of the beams in terms of a significant reduction in the number and width of cracks as well the damage rating [10].

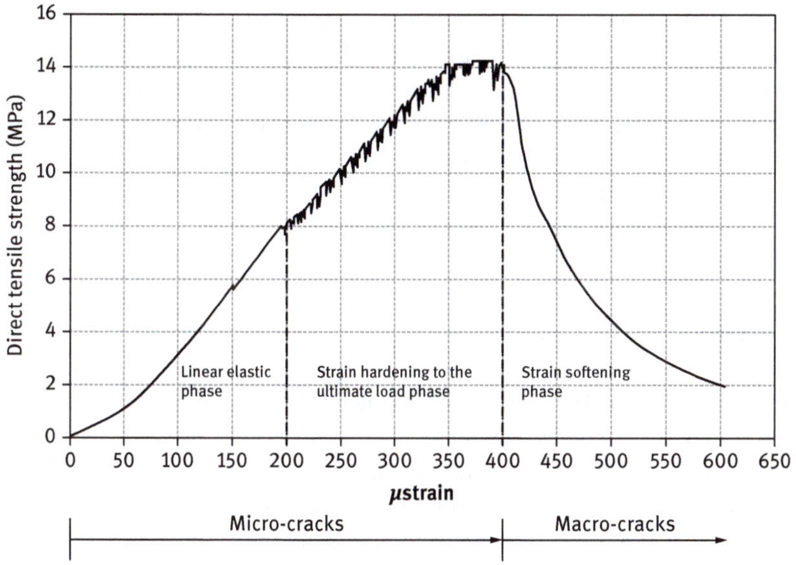

Fig. 5.4: Direct tensile behavior of optimum GUSMRC mix at 90 days [9].

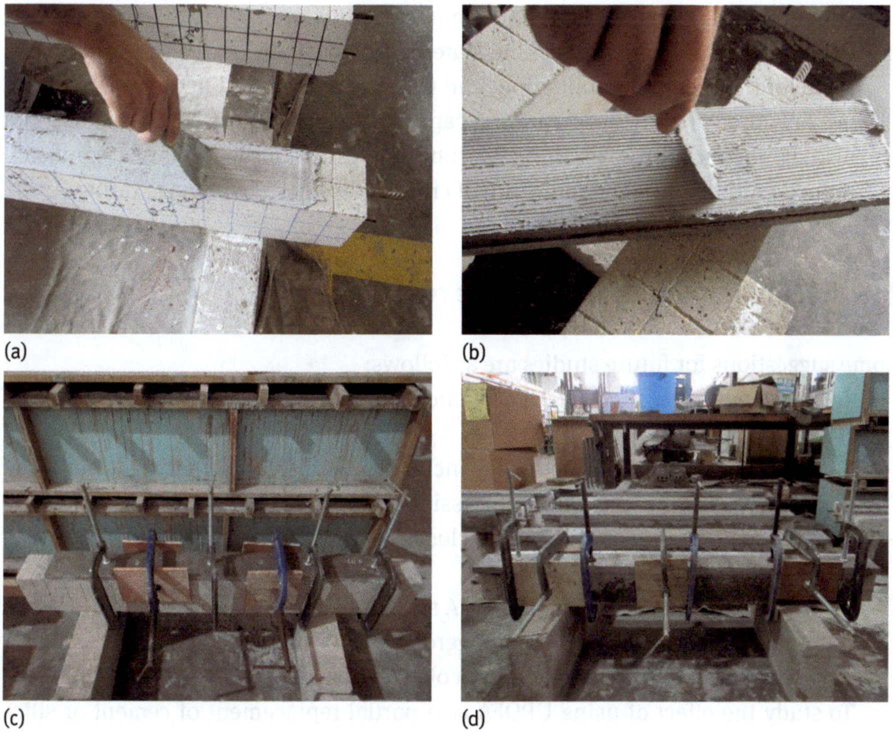

Fig. 5.5: Retrofitting of damaged RC beams using GUSMRC precast strips.

It should be mentioned that several retrofitting configurations have been applied to improve the behavior of damaged RC beams. In addition, the compatibility between GUSMRC strips and the parent materials of beams has been investigated using one of the most popular nondistractive tests, which is the Acoustic Emission (AE) technique. AE was also used for monitoring and assessing the behavior of RC beams before and after retrofitting with GUSMRC. We have conducted extensive testing and analysis on more than 120 beams. Details of these tests will be published soon.

5.5 Conclusion

This present study can contribute immensely to the field of concrete repair and rehabilitation and concrete technology, where:

1. It is shown that the RSM prediction was satisfactory in optimizing the amount of binder contents in the production of traditional UHPFRCC and GUSMRC.
2. UPOFA is an efficient pozzolanic mineral admixture for the production of GUSMRC with possibly superior engineering properties. This could lead to better utilization of POFA in concrete, which subsequently also protects the environment by minimizing the volume of disposed wastes and emissions of greenhouse gases, thereby supporting the sustainability of the concrete industry.
3. GUSMRC is a new green composite material which can be used as retrofitting material for the repair, rehabilitation, and upgrade of damaged RC beams. GUSMRC not only increases the load carrying capacity over that of the control beams, but also significantly improves the serviceability of the beams in terms of a significant reduction in the number and width of cracks as well as the damage rating.

5.6 Recommendations for future research

Some suggestions for future studies are as follows:

- To investigate the effect of UPOFA inclusion on fluid transport properties of GUSMRC.
- To investigate the effect of UPOFA inclusion on the steel fiber distribution of GUSMRC mixtures using image processing techniques.
- To investigate the effect of UPOFA inclusion on the cohesiveness and viscosity of the GUSMRC mixtures.
- To investigate the effect of using UPOFA as a supplementary cementitious material on performance of GUSMRC in an aggressive environment such as sea water or sulfate environment, under long-term observation.
- To study the effect of using UPOFA as a partial replacement of cement or silica fume in combination with other mineral admixtures such as FA, MK, GGBS, etc., on the properties of GUSMRC.

- To improve the mechanical behavior of damaged RC beams caused by corrosion or fire using GUSMRC as retrofitting material.
- To apply GUSMRC as retrofitting material for improving the mechanical behavior of other damaged concrete structural members (e.g., columns, slabs, etc.).

References

[1] Stuart DM. Concrete deterioration. PDH Online Course S155. www.pdhonline.com. Fairfax, VA; 2012.

[2] Pytel A, Kiusalaas J. Engineering mechanics: Statics-SI version. Cengage Learning; 2010.

[3] Alaee F, Karihaloo B. Retrofitting of reinforced concrete beams with CARDIFRC. Journal of Composites for Construction. 2003;7(3):174–186.

[4] Wassermann R, Katz A, Bentur A. Minimum cement content requirements: a must or a myth? Materials and Structures. 2009;42(7):973–982.

[5] Yurdakul E. Optimizing concrete mixtures with minimum cement content for performance and sustainability. PhD thesis, Iowa State University; 2010.

[6] Bucciarelli LL. Ethics and engineering education. European Journal of Engineering Education. 2008;33(2):141–149.

[7] Aldahdooh MAA, Bunnori NM, Megat JMA. Development of green ultra-high performance fiber reinforced concrete containing ultrafine palm oil fuel ash. Construction and Building Materials. 2013;48(0):379–389.

[8] Aldahdooh MAA, Bunnori NM, Megat Johari MA. Evaluation of ultra-high-performance-fiber reinforced concrete binder content using the response surface method. Materials & Design. 2013;52(0):957–965.

[9] Aldahdooh MAA, Bunnori NM, Megat Johari MA. Influence of palm oil fuel ash on ultimate flexural and uniaxial tensile strength of green ultra-high performance fiber reinforced cementitious composites. Materials & Design. 2014;54(0):694–701.

[10] Aldahdooh M, Bunnori NM, Johari MM, Jamrah A, Alnuaimi A. Retrofitting of damaged reinforced concrete beams with a new green cementitious composites material. Composite Structures. 2016;142:27–34.

[11] Spasojević A. Structural implications of ultra-high performance fiber-reinforced concrete in bridge design. PhD thesis, EPFL; 2008.

[12] Russell HG, Graybeal BA. Ultra-high performance concrete: A state-of-the-art report for the bridge community. Georgetown Pike, McLean, VA: Federal Highway Administration; 2013.

[13] Benson S, Karihaloo B. CARDIFRC®-Development and mechanical properties. Part III: Uniaxial tensile response and other mechanical properties. Magazine of Concrete Research. 2005;57(8):433–443.

[14] Richard P, Cheyrezy M. Composition of reactive powder concretes. Cement and Concrete Research. 1995;25(7):1501–1511.

[15] Perry VH, Seibert PJ. The use of UHPFRC (Ductal®) for bridges in North America: The technology, applications and challenges facing commercialization. Proceedings of the Second International Symposium on Ultra High Performance Concrete. Kassel; 2008. p. 815–822.

[16] Park SH, Kim DJ, Ryu GS, Koh KT. Tensile behavior of ultra high performance hybrid fiber reinforced concrete. Cement and Concrete Composites. 2012;34(2):172–184.

[17] Toledo Filho RD, Koenders EAB, Formagini S, Fairbairn EMR. Performance assessment of ultra high performance fiber reinforced cementitious composites in view of sustainability. Materials & Design. 2012;36(0):880–888.

[18] Wille K, Naaman AE, Parra-Montesinos GJ. Ultra-high performance concrete with compressive strength exceeding 150 MPa (22 ksi): a simpler way. ACI Materials Journal. 2011;108(1):46–54.
[19] Benson S, Karihaloo B. CARDIFRC®-Development and mechanical properties. Part I: Development and workability. Magazine of Concrete Research. 2005;57(6):347–352.
[20] Karihaloo BL. CARDIFRC – From concept to industrial application. In: Parra-Montesinos G, Reinhardt H, Naaman AE, editors. High performance fiber reinforced cement composites 6: Springer Netherlands; 2012. p. 397–404.
[21] de Larrard F, Sedran T. Mixture-proportioning of high-performance concrete. Cement and Concrete Research. 2002;32(11):1699–1704.
[22] Graybeal BA. Characterization of the behavior of ultra-high performance concrete. PhD thesis, Georgetown Pike, McLean, VA: McLean University; 2006.
[23] Toutlemonde F, Fouré B, Sorelli L, Baby F, Marchand P, Ulm F. Designing and building with UHPFRC. New York: Wiley Online Library; 2010.
[24] Nematollahi B, Raizal Saifulnaz MR, Jaafar MS, Voo YL. A review on ultra high performance 'ductile' concrete (UHPdC) technology. International Journal of Civil and Structural Engineering. 2012;2(3):994–1009.
[25] Chindaprasirt P, Homwuttiwong S, Jaturapitakkul C. Strength and water permeability of concrete containing palm oil fuel ash and rice husk–bark ash. Construction and Building Materials. 2007;21(7):1492–1499.
[26] Jaturapitakkul C, Kiattikomol K, Tangchirapat W, Saeting T. Evaluation of the sulfate resistance of concrete containing palm oil fuel ash. Construction and Building Materials. 2007;21(7):1399–1405.
[27] Awal A, Hussin MW. The effectiveness of palm oil fuel ash in preventing expansion due to alkali-silica reaction. Cement and Concrete Composites. 1997;19(4):367–372.
[28] MPOB. Malaysian Oil Palm Statistics Malaysia: Economics & Industry Development Division, [Accessed online: 25-2-2014] at http://econ.mpob.gov.my/economy/annual/stat2007/EID_statistics07.htm; 2007.
[29] Al Bawwab WEMK. Probabilistic assessment of liquefaction-induced lateral ground deformations. Middle East Technical University; 2005.
[30] Tangchirapat W, Saeting T, Jaturapitakkul C, Kiattikomol K, Siripanichgorn A. Use of waste ash from palm oil industry in concrete. Waste Management. 2007;27(1):81–88.
[31] Rukzon S, Chindaprasirt P. An experimental investigation of the carbonation of blended portland cement palm oil fuel ash mortar in an indoor environment. Indoor and Built Environment. 2009;18(4):313–318.
[32] Tangchirapat W, Jaturapitakkul C, Chindaprasirt P. Use of palm oil fuel ash as a supplementary cementitious material for producing high-strength concrete. Construction and Building Materials. 2009;23(7):2641–2646.
[33] Megat Johari MA, Zeyad AM, Muhamad Bunnori N, Ariffin KS. Engineering and transport properties of high-strength green concrete containing high volume of ultrafine palm oil fuel ash. Construction and Building Materials. 2012;30(0):281–288.
[34] Tay JH. Ash from oil-palm waste as a concrete material. Journal of Materials in Civil Engineering. 1990;2(2):94–105.
[35] Awal A, Hussin M. Durability of high performance concrete containing palm oil fuel ash. Eighth International Conference on Durability of Building Materials and Components. Vancouver, Canada: National Research Council Canada; 1999. p. 465–474.
[36] Altwair NM, Johari M, Saiyid Hashim S. Flexural performance of green engineered cementitious composites containing high volume of palm oil fuel ash. Construction and Building Materials. 2012;37:518–525.

[37] Ahmad M, Omar R, Malek M, Noor N, Thiruselvam S. Compressive strength of palm oil fuel ash concrete. Proc of the International Conference on Construction and Building Technology. Kuala Lumpur, Malaysia; 2008. p. 297–306.
[38] Karim M, Zain M, Jamil M, Islam M. Strength of concrete as influenced by palm oil fuel ash. Australian Journal of Basic and Applied Sciences. 2011;5(5):990–997.
[39] Sata V, Jaturapitakkul C, Kiattikomol K. Utilization of palm oil fuel ash in high-strength concrete. Journal of Materials in Civil Engineering. 2004;16(6):623–628.
[40] Sata V, Jaturapitakkul C, Kiattikomol K. Influence of pozzolan from various by-product materials on mechanical properties of high-strength concrete. Construction and Building Materials. 2007;21(7):1589–1598.
[41] Auta M, Hameed BH. Optimized waste tea activated carbon for adsorption of methylene blue and acid blue 29 dyes using response surface methodology. Chemical Engineering Journal. 2011;175(0):233–243.
[42] Bashir MJ, Aziz HA, Yusoff MS, Adlan MN. Application of response surface methodology (RSM) for optimization of ammoniacal nitrogen removal from semi-aerobic landfill leachate using ion exchange resin. Desalination. 2010;254(1):154–161.
[43] Ghafari S, Aziz HA, Isa MH, Zinatizadeh AA. Application of response surface methodology (RSM) to optimize coagulation–flocculation treatment of leachate using poly-aluminum chloride (PAC) and alum. Journal of Hazardous Materials. 2009;163(2):650–656.
[44] Baş D, Boyacı İH. Modeling and optimization I: Usability of response surface methodology. Journal of Food Engineering. 2007;78(3):836–845.
[45] ACI-211.1. Standard practice for selecting proportions for normal, heavyweight, and mass concrete. American Concrete Institute (ACI); 1991. p. 211.1–17.

Index

https://doi.org/10.1515/9783110435788-007

Bei Fragen zur Produktsicherheit wenden Sie sich bitte an:
If you have any questions regarding product safety,
please contact:

Walter de Gruyter GmbH
Genthiner Straße 13
10785 Berlin
productsafety@degruyterbrill.com